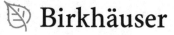

Frontiers in Mathematics

Jihoon Lee • Carlos Morales

Gromov-Hausdorff Stability of Dynamical Systems and Applications to PDEs

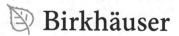 Birkhäuser

Jihoon Lee
Department of Mathematics
Chonnam National University
Gwangju, Korea (Republic of)

Carlos Morales
Mathematics Institute
Federal University of Rio de Janeiro
Rio de Janeiro, Brazil

ISSN 1660-8046   ISSN 1660-8054   (electronic)
Frontiers in Mathematics
ISBN 978-3-031-12030-5   ISBN 978-3-031-12031-2   (eBook)
https://doi.org/10.1007/978-3-031-12031-2

Mathematics Subject Classification: 37B05, 35K57, 53C23

This book is published under the imprint Birkhäuser, www.birkhauser-science.com by the registered company
Springer Nature Switzerland AG
The registered company address is: Gewerbestrasse 11, 6330 Cham, Switzerland

Preface

A common definition of a dynamical system is any phenomenon of nature evolving in "time". One of the most important classes of dynamical systems is described by differential equations or difference equations. We can classify dynamical systems according to discrete or continuous time. Specific questions in the qualitative theory of dynamical systems are the long-term behavior of solutions and the stability problem. A main goal is to study the preservation of geometric structure of solutions under perturbations. For this problem, in early 1960s, Smale used the theory of differential topology to understand the structure of solutions on compact smooth manifolds. One of the most successful notions is hyperbolicity, which has been well developed due to Anosov, Smale, Palis, Mañé, etc. In the light of the rich consequence of differentiable dynamical systems, there are many works that have studied the dynamics properties from a topological viewpoint. For example, instead of hyperbolic systems, it was also shown that any expansive system with shadowing property is topologically stable and admits the spectral decomposition. These results are known as Walters' stability theorem and spectral decomposition theorem, respectively.

One of the important concepts in geometry is the Gromov-Haudorff distance. Roughly, the Gromov-Haudorff distance measures how far two compact metric spaces are from being isometric. Motivated from this, Arbieto and Morales [5] introduced the Gromov-Hausdorff distance between two maps on two compact metric spaces and applied it to study the stability of dynamical systems under perturbations of both maps and phase spaces. We observe that the Gromov-Hausdorff distance is a strong tool to study the stability of dynamical systems.

In the first part of the book, we introduce the notion of Gromov-Hausdorff distance D_{GH} between two dynamical systems and study the stability of dynamical systems under Gromov-Hausdorff perturbations. Note that distance D_{GH} induces a topology on the collection $\mathcal{DS}$ of all dynamical systems up to isometries.

In the second part of the book, we study the stability of dynamical systems induced by dissipative partial differential equations under perturbations of the domain and equation. One of the difficulties in this direction is that the phase space of the induced dynamical system can be changed as we perturb the domain. To overcome this difficulty, we use the Gromov-Hausdorff distance between two dynamical systems.

Part I consists of four chapters to study the abstract Gromov-Hausdorff perturbation theory of dynamical systems. More precisely, we introduce some background concerning the Gromov-Hausdorff space and the Gromov-Hausdorff distance between two dynamical systems in Chap. 1. In Chap. 2, we study a notion of stability, called the topological Gromov-Hausdorff stability, for maps and group actions on compact metric spaces. In Chap. 3, we introduce the notion of orbit shift Gromov-Hausdorff topological stability and study the relationship with the topologically Gromov-Hausdorff stable maps or Anosov maps. In Chap. 4, we analyze the shadowing property of dynamical systems under Gromov-Hausdorff perturbations.

Part II consists of three chapters in which we apply the Gromov-Hausdorff perturbation theory in Part I to study the stability of dynamical systems induced by the dissipative partial differential equations like reaction diffusion equations or Chafee-Infante equations.

Gwangju, Korea (Republic of)   Jihoon Lee
Rio de Janeiro, Brazil   Carlos Morales
October 2022

Contents

Part I
Abstract Theory

Gromov-Hausdorff Distances

1

1.1 Introduction

Metric spaces may be considered as elements in a large space. This was the basis of J. A. Wheeler's suggestion that the dynamic object in Einstein's general relativity is not spacetime, but space. He then defined the *superspace*, namely, the set of isometry classes of Riemannian metrics. He considered this space as the arena where geometrodynamics takes place [77].

In 1975, Edwards [27] defined a distance on Wheeler's superspace which a fortiori extends to the space of isometry classes of compact metric spaces. This distance was rediscovered by Gromov, who used it in his celebrated Polynomial Growth Theorem [31]. This distance is now called *Gromov-Hausdorff distance*.

We start this chapter with a review of the Gromov-Hausdorff distance. It includes not only its precompactness and completeness, but also a variational principle which seems to be new. We also discuss the possible existence of metric space distances characterizing homeomorphic rather than isometric spaces. Afterwards, we extend the Gromov-Hausdorff distance to continuous maps between metric spaces. This leads to the Gromov-Hausdorff space of continuous maps. Finally, we extend this distance to dynamical systems such as flows or finitely generated group actions on metric spaces.

1.2 Preliminary Facts

When two compact metric spaces X and Y are regarded as elements of a large space, it may be of interest to consider possible relations between them. A natural one is as follows. An *isometry* between X and Y is a surjective map $i : X \to Y$ such that $d(i(x), i(x')) = d(x, x')$ for every $x, x' \in X$. We say that X and Y are *isometric* if there is an isometry

© The Author(s), under exclusive license to Springer Nature Switzerland AG 2022
J. Lee, C. Morales, *Gromov-Hausdorff Stability of Dynamical Systems and Applications to PDEs*, Frontiers in Mathematics, https://doi.org/10.1007/978-3-031-12031-2_1

between X and Y. The composition of (two or more) isometries is again an isometry. Therefore, existence of an isometry is an equivalence relation in the space of all compact metric spaces.

Let us introduce the notion of approximate isometry.

Fix $\delta > 0$. A δ-*isometry* is a (not-necessarily continuous) map $i : X \to Y$ such that

$$\sup_{x,x' \in X} |d(i(x), i(x')) - d(x, x')| \le \delta \quad \text{and} \quad d_H(i(X), Y) \le \delta.$$

Here $d_H(A, C)$ denotes the *Hausdorff distance*, defined by

$$d_H(A, C) = \inf\{\varepsilon > 0 : A \subset B(C, \varepsilon) \text{ and } C \subset B(A, \varepsilon)\},$$

where $B(A, \varepsilon) = \bigcup_{a \in A} B(a, \varepsilon)$ is the ε-ball centered at A. The *diameter* of a nonempty subset $A \subset X$ is defined as

$$\text{diam}(A) = \sup_{a,a' \in A} d(a, a').$$

The composition of an isometry and a δ-isometry is considered in the next lemma.

Lemma 1.1 *If* $i : X \to Y$ *is a* δ-*isometry (resp., an isometry) and* $j : Y \to Z$ *is an isometry (resp., a* δ-*isometry), then* $j \circ i : X \to Z$ *is a* δ-*isometry.*

Proof Let us consider first the case when i and j are a δ-isometry and an isometry, respectively. For $x, x' \in X$, we have

$$|d(j \circ i(x), j \circ i(x')) - d(x, x')| = |d(i(x), i(x')) - d(x, x')| \le \delta,$$

and so

$$\sup_{x,x' \in X} |d(j \circ i(x), j \circ i(x')) - d(x, x')| \le \delta.$$

On the other hand, if $z \in Z$, then $j(y) = z$ for some $y \in Y$. Moreover, there is $x \in X$ such that $d(i(x), y) \le \delta$. Then $d(j \circ i(x), z) = d(j(i(x)), j(y)) = d(i(x), y) \le \delta$, proving that $d_H(j \circ i(X), Z) \le \delta$. We conclude that $j \circ i$ is a δ-isometry.

Now suppose that i and j are an isometry and a δ-isometry, respectively. Then $|d(j \circ i(x), j \circ i(x')) - d(x, x')| = |d(j(i(x)), j(i(x'))) - d(i(x), i(x'))| \le \delta$ for $x, x' \in X$, and so

$$\sup_{x,x' \in X} |d(j \circ i(x), j \circ i(x')) - d(x, x')| \le \delta.$$

If $z \in Z$, then $d(j(y), z) \le \delta$ for some $y \in Y$ and $y = i(x)$ for some $x \in X$. So, $d(j \circ i(x), z) = d(j(y), z) \le \delta$. Therefore, $d_H(j \circ i(X), Z) \le \delta$, completing the proof.

$\square$

We now consider the composition of δ-isometries.

Proposition 1.1 *If $i : X \to Y$ and $j : Y \to Z$ are a δ_1-isometry and a δ_2-isometry respectively, of compact metric spaces, then $j \circ i : X \to Z$ is a $2(\delta_1 + \delta_2)$-isometry.*

Proof For all $x, x' \in X$, we have

$$d(j \circ i(x), j \circ i(x')) \le \delta_2 + d(i(x), i(x')) \le (\delta_1 + \delta_2) + d(x, x')$$

and

$$d(x, x') \le \delta_1 + d(i(x), i(x')) \le (\delta_1 + \delta_2) + d(j \circ i(x), j \circ i(x')).$$

Consequently,

$$\sup_{x, x' \in X} |d(j \circ i(x), j \circ i(x')) - d(x, x')| \le \delta_1 + \delta_2.$$

Also, if $z \in Z$, then there is an $y \in Y$ such that $d(j(y), z) \le \delta_2$; and for this y, there is an $x \in X$ such that $d(i(x), y) \le \delta_1$. Then,

$$d(j \circ i(x), z) \le d(j(i(x)), j(y)) + d(j(y), z)$$
$$\le 2\delta_2 + d(i(x), y)$$
$$\le \delta_1 + 2\delta_2$$
$$\le 2(\delta_1 + \delta_2).$$

Therefore, $d_H(j \circ i(X), Z) \le 2(\delta_1 + \delta_2)$, and we conclude that $j \circ i$ is a $2(\delta_1 + \delta_2)$-isometry from X to Z.

$\square$

The following theorem is the basis of the definition of the Gromov-Hausdorff distance. It just characterizes isometric spaces in terms of δ-isometries.

Theorem 1.1 *Two compact metric spaces X and Y are isometric if and only if*

$$\inf\{\Delta > 0 : \text{there is a } \Delta\text{-isometry } i : X \to Y\} = 0.$$

Proof The sufficiency follows from the obvious fact that every isometry is a Δ-isometry for every $\Delta > 0$. For the necessity, suppose that the infimum in the statement is zero. Then there is a sequence of Δ_n-isometries $\{i_n : X \to Y\}_{n \in \mathbb{N}}$ with $\Delta_n \to 0$ as $n \to \infty$. Since X is compact, there exists a countable dense subset A of X. Since Y is compact, the sequence $\{i_n(a)\}_{n \in \mathbb{N}}$ has a convergent subsequence for every $a \in A$. Since A is countable, by a diagonal argument, there is a subsequence $\{i_{n_k}\}_{k \in \mathbb{N}}$ such that $(i_{n_k}(a))_{k \in \mathbb{N}}$ is convergent for every $a \in A$. Hence, there exists a map $i : A \to Y$ such that the sequence $i_{n_k} : A \to Y$ converges pointwise to i. Since

$$d(i(a), i(a')) = \lim_{k \to \infty} d(i_{n_k}(a), i_{n_k}(a')) \leq \lim_{k \to \infty} (\Delta_{n_k} + d(a, a')) = d(a, a')$$

for every $a, a' \in A$, we conclude that i is uniformly continuous. Hence, since A is dense, $i : A \to Y$ admits a continuous extension, still denoted by $i : X \to Y$, such that $d(i(x), i(x')) \leq d(x, x')$ for every $x, x' \in X$.

Now take $y \in X$. Then there is $x_{n_k} \in X$ such that $d(i_{n_k}(x_{n_k}), y) \leq \delta_{n_k}$ for every $k \in \mathbb{N}$. Since X is compact, we can assume that $x_{n_k} \to x$ for some $x \in X$. Fix $\Delta > 0$ and take $a \in A$ such that $d(x, a) < \frac{\Delta}{2}$. We can choose k such that $d(x_{n_k}, a) < \frac{\Delta}{2}$. Then

$$d(i(x), y) \leq d(i(x), i(a)) + d(i(a), i_{n_k}(a)) + d(i_{n_k}(a), i_{n_k}(x_{n_k})) + d(i_{n_k}(x_{n_k}), y),$$

and so

$$d(i(x), y) < \frac{\Delta}{2} + 2\delta_{n_k} + \frac{\Delta}{2} = \Delta + 2\delta_{n_k} \to \Delta$$

as $k \to \infty$, proving that $d(i(x), y) \leq \Delta$. If follows that there exists a sequence $\{x_n\}_{n \in \mathbb{N}}$ in X such that $d(i(x_n), y) \to 0$ as $n \to \infty$, and since X is compact, we can assume $x_n \to x$ for some $x \in X$. Hence, $i(x) = y$, proving that i is surjective. Thus, $i : X \to Y$ is a *surjective* map between compact metric spaces such that $d(i(x), i(x')) \leq d(x, x')$ for every $x, x' \in X$. This implies that i is an isometry (Exercise 1.15), which completes the proof.   □

1.3   Gromov-Hausdorff Distance for Metric Spaces

An important structure a space may be equipped with is that of a metric or distance. Indeed, any set can be endowed with one, i.e., the discrete metric. The full object of study in this section will be the set of all compact metric spaces. We will search a distance in this space which differentiates isometric spaces. The search for distances that differentiate homeomorphic spaces instead will be carried out in Sect. 1.6.

Theorem 1.1 presents a necessary and sufficient condition for two compact metric spaces X and Y to be isometric. Namely, when the infimum of those $\Delta > 0$ for which there is a Δ-isometry $i : X \to Y$ is zero. Therefore, thus infimum represents a measure of how far X and Y are from being isometric.

Definition 1.1 The *Gromov-Hausdorff distance* between two compact metric spaces X and Y is defined by

$$d_{\mathrm{GH}}(X, Y) = \inf\{\Delta > 0 : \text{there are } \Delta\text{-isometries } i : X \to Y \text{ and } j : Y \to X\}.$$

The following theorem summarizes the main properties of d_{GH}.

Theorem 1.2 *The following properties hold for every triple of compact metric spaces X, Y, and Z:*

(1) $0 \leq d_{\mathrm{GH}}(X, Y) < \infty$ *and* $d_{\mathrm{GH}}(X, Y) = 0$ *if and only if X and Y are isometric;*
(2) $d_{\mathrm{GH}}(X, Y) = d_{\mathrm{GH}}(Y, X)$; *and*
(3) $d_{\mathrm{GH}}(X, Z) \leq 2(d_{\mathrm{GH}}(X, Y) + d_{\mathrm{GH}}(Y, Z))$.

Proof First, we prove Item (1). Clearly, $d_{\mathrm{GH}}(X, Y) \geq 0$. Fix $y_0 \in Y$ and define $i : X \to Y$ by $i(x) = y_0$ for every $x \in X$. Since

$$|d(i(x), i(x')) - d(x, x')| = d(x, x') \leq \mathrm{diam}(X)$$

for $x, x' \in X$, we have

$$\sup_{x,x' \in X} |d(i(x), i(x')) - d(x, x')| \leq \mathrm{diam}(X).$$

Also, $d(i(x), y) = d(y_0, y) \leq \mathrm{diam}(Y)$ for all $y \in Y$. So, $d_{\mathrm{H}}(i(X), Y) \leq \mathrm{diam}(Y)$. All together, these facts imply that i is a $\max\{\mathrm{diam}(X), \mathrm{diam}(Y)\}$-isometry from X to Y. Likewise, fixing $x_0 \in X$ and defining $j : Y \to X$ by $j(y) = x_0$ for every $y \in Y$, we obtain a $\max\{\mathrm{diam}(X), \mathrm{diam}(Y)\}$-isometry from Y to X. Thus shows that $d_{\mathrm{GH}}(X, Y) \leq \max\{\mathrm{diam}(X), \mathrm{diam}(Y)\} < \infty$. The fact that $d_{\mathrm{GH}}(X, Y) = 0$ if and only if X and Y are isometric is a direct consequence of Theorem 1.1. This proves Item (1).

Item (2) is trivial (which is why the Δ-isometry $j : Y \to X$ in the definition of d_{GH} was added).

Finally, we prove Item (3). Fix $\delta > 0$. It follows from the definition of d_{GH} that there are $(d_{\mathrm{GH}}(X, Y) + \delta)$-isometries $i : X \to Y$ and $j : Y \to X$; and $(d_{\mathrm{GH}}(Y, Z) + \delta)$-isometries $k : Y \to Z$ and $l : Z \to Y$. By Proposition 1.1, $k \circ i : X \to Z$ and $j \circ l : Z \to X$ are $2(d_{\mathrm{GH}}(X, Y) + d_{\mathrm{GH}}(Y, Z) + 2\delta)$-isometries. Therefore, $d_{\mathrm{GH}}(X, Z) \leq 2(d_{\mathrm{GH}}(X, Y) + d_{\mathrm{GH}}(Y, Z)) + 4\delta$. Since δ is arbitrary, we get Item (3). □

Hereafter we will use the following definition

$$M = \{X : X \text{ is a compact metric space}\}.$$

We write $X \approx Y$ for $X, Y \in M$ if X and Y are isometric.

Lemma 1.2 *If $X \approx X'$ and $Y \approx Y'$ for $X, X', Y, Y' \in M$, then*

$$d_{GH}(X, Y) = d_{GH}(X', Y').$$

Proof By hypothesis, there are isometries $A : X \to X'$ and $B : Y \to Y'$. Fix $\delta > 0$. By definition, there are $(d_{GH}(X, Y) + \delta)$-isometries $i : X \to Y$ and $j : Y \to X$. By Lemma 1.1, $B \circ i : X \to Y'$ and $j \circ B^{-1} : Y' \to X$ are $(d_{GH}(X, Y)+\delta)$-isometries, whence $d_{GH}(X, Y') \leq d_{GH}(X, Y) + \delta$. Since δ is arbitrary, $d_{GH}(X, Y') \leq d_{GH}(X, Y)$. Reversing the roles of Y and Y', we obtain $d_{GH}(X, Y) \leq d_{GH}(X, Y')$. Therefore, $d_{GH}(X, Y) = d_{GH}(X, Y')$. It follows that $d_{GH}(X, Y) = d_{GH}(X, Y') = d_{GH}(X', Y')$. □

As we already mentioned, $\approx$ is an equivalence relation of M. We denote

$$\mathcal{M} = M/\approx$$

the set of equivalence classes of $\approx$ and denote by $[X]$ the equivalence class of $X \in M$.

Lemma 1.2 implies that the map $d_{GH} : \mathcal{M} \times \mathcal{M} \to [0, \infty[$ given by

$$d_{GH}([X], [Y]) = d_{GH}(X, Y), \quad [X], [Y] \in \mathcal{M},$$

is well-defined. Hereafter, in order to simplify notations, we will still denote by X the elements of $\mathcal{M}$, understanding that X is an equivalence class rather than a single metric space.

Theorem 1.2 implies the following properties:

1. $0 \leq d_{GH}(X, Y) < \infty$ and $d_{GH}(X, Y) = 0$ if and only if $X = Y$,
2. $d_{GH}(X, Y) = d_{GH}(Y, X)$; and
3. $d_{GH}(X, Z) \leq 2(d_{GH}(X, Y) + d_{GH}(Y, Z))$ for every $X, Y, Z \in \mathcal{M}$.

This means that $(\mathcal{M}, d_{GH})$ is a so-called *quasi-metric space*.

Definition 1.2 The space $\mathcal{M} = (\mathcal{M}, d_{GH})$ is called the *Gromov-Hausdorff space*.

This space will be object of study in the next sections.

1.4   Precompactness, Completeness, and a Variational Principle

Many authors have studied the topology of the Gromov-Hausdorff space $\mathcal{M}$. It is not difficult to see that $\mathcal{M}$ is separable, since the set of finite metric spaces with rational-valued distance functions is dense in $\mathcal{M}$. We can also prove that $\mathcal{M}$ is complete. Indeed, we will obtain this as a corollary of the so-called *Gromov precompactness theorem* stated as follows.

Given $\varepsilon > 0$, an *ε-net* of a metric space X is a subset $S \subset X$ such that $d_H(S, X) < \varepsilon$, i.e., for every $x \in X$, there is $s \in S$ such that $d(s, x) < \varepsilon$. The cardinality of S is denoted by $|S|$. A subset F of a metric space X is *precompact* if the closure of F in X is compact and, equivalently, if every sequence in F has a convergent subsequence.

Theorem 1.3 *A subset $\mathcal{X}$ of $\mathcal{M}$ is precompact if and only if the following properties hold:*

(1) $\sup_{X \in \mathcal{X}} \operatorname{diam}(X) < \infty$;
(2) *For every $\varepsilon > 0$, there is $N(\varepsilon) \in \mathbb{N}$ such that every $X \in \mathcal{X}$ has an ε-net with at most $N(\varepsilon)$ elements.*

Proof We will prove the sufficiency and leave the necessity as an exercise (see Exercise 1.20). Let $\{X_n\}_{n \in \mathbb{N}}$ be a sequence in $\mathcal{X}$. We need to prove that this sequence has a convergent subsequence. This is done as follows.

First, take $D = \sup_{X \in \mathcal{X}} \operatorname{diam}(X)$. Then $D < \infty$ by Item (1).

Next, take an arbitrary countable set $X = \{x_1, x_2, \ldots\}$. We will define a map $d : X \times X \to [0, \infty[$ through the following procedure. Take $\varepsilon = \frac{1}{k}$ for $k \in \mathbb{N}$ in Item (2) to obtain the sequence $\{N(\frac{1}{k})\}_{k \in \mathbb{N}}$. It follows from Item (2) that given $n \in \mathbb{N}$, there is a sequence $E_1, E_2, \ldots, E_k, \ldots \subset X_n$ such that E_k is a $\frac{1}{k}$-net of X_n with at most $N(\frac{1}{k})$ elements. Then we can organize $S_n = \bigcup_{k \geq 1} E_k = (x_i^n)_{i \in \mathbb{N}}$ so that the first N_k elements form a $\frac{1}{k}$-net of X_n, where $(N_k)_{k \in \mathbb{N}}$ is defined by $N_1 = N(1)$ and $N_k = N_{k-1} + N(\frac{1}{k})$ for $k \geq 2$.

By using Item (1) in the statement of the theorem, we have that $\{d(x_i^n, x_j^n) : n, i, j \in \mathbb{N}\}$ is bounded (in fact, contained in $[0, D]$). Then a diagonal argument allows us to choose a subsequence $\{X_{n_k}\}_{k \in \mathbb{N}}$ of $\{X_n\}_{n \in \mathbb{N}}$ such that $(d(x_i^{n_k}, x_j^{n_k}))_{k \in \mathbb{N}}$ converges for every $i, j \in \mathbb{N}$. For simplicity, we assume that $n_k = n$ for every $k \in \mathbb{N}$. Hence, the real number sequence $\{d(x_i^n, x_j^n)\}_{n \in \mathbb{N}}$ converges for every $i, j \in \mathbb{N}$.

Finally, we define $d : X \times X \to [0, \infty[$ by

$$d(x_i, x_j) = \lim_{n \to \infty} d(x_i^n, x_j^n)$$

for all $i, j \in \mathbb{N}$. Thus d is non-negative, finite, symmetric, and satisfies the triangle inequality, but may be zero at different points x_i, x_j. By identifying such points, we obtain a metric space still denoted by (X, d). Denote by X_∞ the completion of (X, d). To finish the proof, we shall prove the following properties.

- X_∞ *is compact.*

Since X_∞ is complete, it suffices to prove that X_∞ has a finite $\frac{1}{k}$-net for all $k \in \mathbb{N}$ (Exercise 1.16).

First, we prove that $S^{(k)} = \{x_1, x_2, \ldots, x_{N_k}\}$ is a $\frac{1}{k}$-net of X. Take any $x_i \in X$. It follows from the choices that $S_n^{(k)} = \{x_i^n : 1 \leq i \leq N_k\}$ is a $\frac{1}{k}$-net of X_n. So, there is a sequence $\{j_n\}_{n \in \mathbb{N}}$ with $1 \leq j_n \leq N_k$ such that $d(x_i^n, x_{j_n}^n) \leq \frac{1}{k}$ for every n. Since N_k is finite and does not depend on n, there is $1 \leq j \leq N_k$ such that $j_n = j$ for infinitely many n's. It follows that $d(x_i^n, x_j^n) \leq \frac{1}{k}$ for infinitely many n's. By taking $n \to \infty$, we get $d(x_i, x_j) \leq \frac{1}{k}$, concluding that $S^{(k)}$ is a $\frac{1}{k}$-net of X.

Finally, since X is dense in X_∞, $S^{(k)}$ is also a $\frac{1}{k}$-net of X_∞. Since $S^{(k)}$ is finite, we are done.

- $X_n \to X_\infty$.

We have

$$d_{\mathrm{GH}}(X_n, X_\infty) \leq 2(d_{\mathrm{GH}}(X_n, S_n^{(k)}) + d_{\mathrm{GH}}(S_n^{(k)}, X_\infty)),$$

and so

$$d_{\mathrm{GH}}(X_n, X_\infty) \leq 2\left(\frac{1}{k} + 2\left(d_{GH}(S_n^{(k)}, S^{(k)}) + \frac{1}{k}\right)\right).$$

Since $d_{\mathrm{GH}}(S_n^{(k)}, S^{(k)}) \to 0$ as $n \to \infty$, we get

$$\lim_{n \to \infty} d_{\mathrm{GH}}(X_n, X_\infty) \leq \frac{6}{k}$$

for all $k \in \mathbb{N}$, proving the result. $\square$

Recall that a metric space X is *complete* if every Cauchy sequence in X is convergent (a *Cauchy sequence*, is a sequence $\{x_n\}_{n \in \mathbb{N}}$ such that for every $\varepsilon > 0$, there is $N \in \mathbb{N}$ such that $d(x_n, x_m) \leq \varepsilon$ for all $n, m \geq N$).

We will prove that $\mathcal{M}$ is complete. For this, we need two additional lemmas.

Lemma 1.3 *If $S \subset X$ is an ε-net of a compact metric space X and $i : X \to Y$ is a δ-isometry of compact metric spaces, then $i(S)$ is an $(\varepsilon + 2\delta)$-net of Y.*

Proof If $y \in Y$, then there is an $x \in X$ such that $d(i(x), y) \leq \delta$. For this x, we have an $s \in S$ such that $d(x, s) \leq \varepsilon$. Then

$$d(i(s), y) \leq d(i(s), i(x)) + d(i(x), y) \leq \delta + d(x, s) + d(i(x), y) \leq 2\delta + \varepsilon,$$

proving the result. $\square$

Lemma 1.4 *If $\{X_n\}_{n\in\mathbb{N}}$ is a Cauchy sequence in $\mathcal{M}$ and $d_{GH}(X_n, X_{n+1}) < 2^{-(n+1)}$ for $n \in \mathbb{N}$, then*

$$\sup_{n\in\mathbb{N}} \operatorname{diam}(X_n) < \infty.$$

Proof By hypothesis, there is a sequence of $2^{-(n+1)}$-isometries $\{i_n : X_{n+1} \to X_n\}_{n\in\mathbb{N}}$. It follows that

$$d(x_{n+1}, x'_{n+1}) < 2^{-(n+1)} + d(i_n(x_{n+1}), i_n(x'_{n+1})) \le 2^{-(n+1)} + \operatorname{diam}(X_n)$$

for every $x_{n+1}, x'_{n+1} \in X_{n+1}$, whence $\operatorname{diam}(X_{n+1}) \le 2^{-(n+1)} + \operatorname{diam}(X_n)$ for every $n \in \mathbb{N}$. Inductively, we obtain

$$\operatorname{diam}(X_n) \le \operatorname{diam}(X_1) + \sum_{k=1}^{\infty} 2^{-k}$$

for every $n \in \mathbb{N}$. Therefore,

$$\sup_{n\in\mathbb{N}} \operatorname{diam}(X_n) \le \operatorname{diam}(X_1) + \sum_{k=1}^{\infty} 2^{-k} < \infty,$$

proving the result. $\qquad\square$

Now we can prove the following theorem.

Theorem 1.4 *The Gromov-Hausdorff space $\mathcal{M}$ is complete.*

Proof Let $\{X_n\}_{n\in\mathbb{N}}$ be a Cauchy sequence of $\mathcal{M}$. By passing to a subsequence if necessary, we can assume that $d_{GH}(X_n, X_{n+1}) < 2^{-(n+1)}$ for every $n \in \mathbb{N}$. We shall prove that the family $\mathcal{X} = \{X_1, X_2, \ldots\}$ satisfies the conditions of the Gromov precompactness theorem.

First, we observe that $\sup_{n\in\mathbb{N}} \operatorname{diam}(X_n) < \infty$ by Lemma 1.4, and so Item (1) of Theorem 1.3 holds.

Next, we prove Item (2) of Theorem 1.3. Fix $\varepsilon > 0$ and take $n \in \mathbb{N}$ such that

$$\sum_{r=n}^{\infty} 2^{-r} < \frac{\varepsilon}{2}.$$

Since X_k is compact for every $1 \le k \le n$, there is a finite $\frac{\varepsilon}{2}$-net S_k of X_k for all such k. Take $N(\varepsilon) = \max_{1\le k\le n} |S_k|$. Since $d_{GH}(X_k, X_{k+1}) < 2^{-(k+1)}$ for all $k \in \mathbb{N}$, there is a $2^{-(k+1)}$-isometry $i_k : X_k \to X_{k+1}$ for every $k \in \mathbb{N}$. Define

$$S_k = i_k \circ i_{k-1} \circ \cdots \circ i_n(S_n)$$

for every $k > n$. We shall prove that S_k is an ε-net with no more than $N(\varepsilon)$ elements for every $k \in \mathbb{N}$. This is obviously true for $1 \le k \le n$, and so we can assume $k > n$.

For this, we claim that S_k is an $(\frac{\varepsilon}{2} + \sum_{r=n}^{k} 2^{-r})$-net of X_k for every $k > n$. Since S_n is an $\frac{\varepsilon}{2}$-net of X_n, $S_{n+1} = i_n(S_n)$ is an $(\frac{\varepsilon}{2} + 2 \cdot 2^{-(n+1)})$-net (and so an $(\frac{\varepsilon}{2} + 2^{-n})$-net) of X_{n+1} by Lemma 1.2. Then the assertion holds for $k = n + 1$. Now suppose that the assertion is true for $k > n$. Then by Lemma 1.2, we have that $S_{k+1} = i_{k+1}(S_k)$ is an $(\frac{\varepsilon}{2} + \sum_{r=n}^{k} 2^{-r} + 2 \cdot 2^{-(k+2)})$-net (and so an $(\frac{\varepsilon}{2} + \sum_{r=n}^{k+1} 2^{-r})$-net) of X_{k+1} and the claim follows by induction.

Since $|S_k| \le |S_n| \le N(\varepsilon)$, we conclude that S_k is an ε-net with no more than $N(\varepsilon)$ elements for every $k \in \mathbb{N}$.

Then Item (2) holds and so $\mathcal{X}$ is precompact by Theorem 1.3. It follows that $\{X_n\}_{n \in \mathbb{N}}$ has a convergent subsequence. Since $\{X_n\}_{n \in \mathbb{N}}$ is Cauchy, $\{X_n\}_{n \in \mathbb{N}}$ itself is convergent (see Exercise 1.16) and the result follows. $\square$

1.5   A Variational Principle for the Gromov-Hausdorff Distance

In this section we express the Gromov-Hausdorff distance as the solution of an optimization problem in the space of probability measures.

Let X, Y be compact metric spaces, let $\Delta > 0$, and let ν be a Borel probability measure on Y. We say that $i : X \to Y$ is a (ν, Δ)-isometry if

- $|d(i(x), i(x')) - d(x, x')| < \Delta$ for all $x, x' \in X$, and
- $\nu((\mathrm{Im}(i))^{\Delta}) = 1$.

We give two examples.

Example 1.3

(a) $i : X \to Y$ is a Δ-isometry if and only if i is a (ν, Δ)-isometry for every Borel probability measure ν on Y. Indeed, the direct implication is obvious, while the opposite one follows by taking $\nu = \delta_{y_0}$ for $y_0 \in Y$, where δ_{y_0} is the Dirac measure at y_0. Then $y_0 \in (\mathrm{Im}(i))^{\Delta}$ for all $y_0 \in Y$. Hence, we get $Y = (\mathrm{Im}(i))^{\Delta}$.

(b) $i : X \to Y$ is (ν, Δ)-isometry for every $\Delta > 0$ (ν is fixed) if and only if i is an isometric immersion such that

$$\mathrm{Im}(i) \supseteq \mathrm{supp}(\nu).$$

The second example above suggests the following definition.

Definition 1.4 Let X, Y be compact metric spaces and $Y_0 \subseteq Y$. We say that $i : X \to Y$ is a (Δ, Y_0)-isometry if

- $|d(i(x), i(x')) - d(x, x')| < \Delta$ for all $x, x' \in X$, and
- $(\text{Im}(i))^\Delta \supseteq Y_0$.

Definition 1.5 Given the pair (X, X_0) and (Y, Y_0) of compact metric spaces X, Y and subsets $X_0 \subseteq X$ and $Y_0 \subseteq Y$, we define a distance $d((X, X_0), (Y, Y_0))$ as the infimum of the numbers $\Delta > 0$ such that there are a (Δ, Y_0)-isometry $i : X \to Y$ and a (Δ, X_0)-isometry $j : Y \to X$.

A priori, there is no comparison between $d((X, X_0), (Y, Y_0))$ and $d_{\text{GH}}(X, Y)$. However, we have the following result.

Theorem 1.5 *For every pair (X, X_0) and (Y, Y_0), we have*

$$d((X, X_0), (Y, Y_0)) = 0 \quad \textit{if and only if} \quad d_{\text{GH}}(X, Y) = 0.$$

Proof Suppose $d_{\text{GH}}(X, Y) = 0$, which hold if and only if X and Y are isometric. Then for any $\Delta > 0$, there exist Δ-isometries (actually, isometries) $i : X \to Y$ and $j : Y \to X$. So, $i(X) \supseteq Y_0$ and $j(Y) \supseteq X_0$; hence, $d((X, X_0), (Y, Y_0)) = 0$. Conversely, if $d((X, X_0), (Y, Y_0)) = 0$, then by the standard procedure, there are isometric immersions $i : X \to Y$ and $j : Y \to X$. The composition $j \circ i : X \to X$ is an isometric immersion with X compact, so it is onto. Then if $y \in Y$ we have that $f(y) \in X$. Consequently, $j(y) = ji(x)$ for some $x \in X$. Hence, $y = i(x)$ (since j is injective), proving that i is onto and so an isometry. Then X and Y are isometric, proving $d_{\text{GH}}(X, Y) = 0$. $\square$

By a measure metric space (X, ν), we mean a compact metric space X equipped with a Borel measure ν. The following will be the distance between measure metric spaces we will deal with.

Definition 1.6 Given two measure metric spaces (X, ν) and (Y, μ), we define $d((X, \nu), (Y, \mu))$ as the infimum of $\Delta > 0$ such that there are a (ν, Δ)-isometry $i : X \to Y$ and a (μ, Δ)-isometry $j : Y \to X$.

We denote by $\mathbb{P}(X)$ the set of Borel probability measures on X. The following variational principle for the Gromov-Hausdorff distance is obtained.

Theorem 1.6 *If X and Y are compact metric spaces without isolated points, then*

$$d_{\text{GH}}(X, Y) = \sup_{(\nu, \mu) \in \mathbb{P}(X) \times \mathbb{P}(Y)} d((X, \nu), (Y, \mu)).$$

Proof Fix $\varepsilon > 0$ and define

$$\Delta = d_{\mathrm{GH}}(X, Y) + \varepsilon.$$

Then there exist Δ-isometries $i : X \to Y$ and $j : Y \to X$. By Item (2) of Example 1.3, we have that i is a (Δ, ν)-isometry and j is a (Δ, μ)-isometry (for any ν, μ). Then

$$d((X, \nu), (Y, \mu)) < \Delta.$$

Thus,

$$\sup_{(\nu,\mu)\in\mathbb{P}(X)\times\mathbb{P}(Y)} d((X, \nu), (Y, \mu)) \le \Delta = d_{\mathrm{GH}}(x, y) + \varepsilon.$$

Since $\varepsilon > 0$ is arbitrary, we get

$$\sup_{(\nu,\mu)\in\mathbb{P}(X)\times\mathbb{P}(Y)} d((X, \nu), (Y, \mu)) \le d_{\mathrm{GH}}(x, y).$$

For the converse inequality, we deduce from the assumption of nonexistence of isolated points that there exists $(\nu_0, \mu_0) \in \mathbb{P}(X)\times\mathbb{P}(y)$ such that $\mathrm{supp}(\nu_0) = X$ and $\mathrm{supp}(\mu_0) = Y$. By taking $\varepsilon > 0$, $\Delta = d((X, \nu_0), (Y, \mu_0)) + \varepsilon$, and $\Delta' = d((X, \nu_0), (Y, \mu_0)) + 2\varepsilon$, we get a (Δ, μ_0)-isometry $i : X \to Y$ and a (Δ, ν_0)-isometry $j : Y \to X$. But then $\mu_0(\mathrm{Im}(i)^\Delta) = 1$, so $Y = \mathrm{supp}(\mu_0) \subseteq (\mathrm{Im}(i))^{\Delta'}$. Similarly, $X \subseteq (\mathrm{Im}(i))^{\Delta'}$. So, i and j are Δ-isometries, implying that

$$d_{\mathrm{GH}}(X, Y) \le \Delta' = d((X, \nu_0), (Y, \mu_0)) + 2\varepsilon.$$

Since ε is arbitrary, we get

$$d_{\mathrm{GH}}(X, Y) \le d((X, \nu_0), (Y, \mu_0)).$$

Thus,

$$d_{\mathrm{GH}}(x, y) \le \sup_{(\nu,\mu)\in\mathbb{P}(X)\times\mathbb{P}(y)} d((X, \nu), (Y, \mu)).$$

This completes the proof.   □

1.6 Distances Characterizing Homeomorphic Spaces

This section deals with the problem of whether the Gromov-Hausdorff distance can be extended to some distance that is able to detect when two metric spaces are homeomorphic rather than isomorphic. This problem is quite natural and certainly familiar to some authors (though as far as we know nobody wrote about this before). Although in general the answer is negative, the readers (and of course ourselves) are encouraged to overcome the obstacles we found here.

To start, let us recall the definition

$$M = \{X : X \text{ is a compact metric space}\}.$$

Two spaces $X, Y \in M$ are *homeomorphic* if there is a homeomorphism from X to Y. Like isomorphism, homeomorphism is an equivalence relation in M.

Motivated by the Gromov-Hausdorff distance in M, it is natural to study non-zero "distances" in M that vanish on pairs of homeomorphic rather than isomorphic spaces. More precisely, we study non-zero maps $D : M \times M \to [0, \infty[$ satisfying the following properties:

(P1) If X and Y are homeomorphic, then $D(X, Y) = 0$.
(P2) $D(X, Y) = D(Y, X)$ for every $X, Y \in M$.
(P3) There is $K > 0$ such that $D(X, Z) \leq K(D(X, Y) + D(Y, Z))$ for every $X, Y, Z \in M$.

We will give some negative answers with the following definition. Hereafter, $\{*\}$ will denote the one-point set.

Definition 1.7 We say that $D : M \times M \to [0, \infty[$ is *normalized* if there is $L > 0$ such that

$$L^{-1}D(X, \{*\}) \leq \text{diam}(X) \leq LD(X, \{*\}), \qquad \forall X \in M.$$

We can give two examples. The first is precisely the Gromov-Hausdorff distance.

Example 1.8 $d_{\text{GH}}(X, \{*\}) = \text{diam}(X)$ for every $X \in M$. In particular, the Gromov-Hausdorff distance d_{GH} is normalized.

Proof Take $\varepsilon > 0$ and $\Delta < d_{\text{GH}}(X, \{*\}) + \varepsilon$ such that there is a Δ-isometry $i : X \to \{*\}$. Then

$$d(x, x') = |d(*, *) - d(x, x')| = |d(i(x), i(x')) - d(x, x')| \leq \Delta$$

for every $x, x' \in X$, proving $\mathrm{diam}(X) \le \Delta < d_{\mathrm{GH}}(X, \{*\}) + \varepsilon$. Since ε is arbitrary, $\mathrm{diam}(X) \le d_{\mathrm{GH}}(X, \{*\})$. Now suppose that $d_{\mathrm{GH}}(X, \{*\}) < \mathrm{diam}(X)$. Then there exist a $\Delta < \mathrm{diam}(X)$ and a Δ-isometry $i : X \to \{*\}$. Again, we have $d(x, x') \le \Delta$ for every $x, x' \in X$. Thus, $\mathrm{diam}(X) \le \Delta < \mathrm{diam}(X)$, a contradiction. This completes the proof. □

The second is as follows. Define the C^0 distance between maps $r, l : X \to Y$ by

$$d(r, l) = \sup_{x \in X} d(r(x), l(x)).$$

Given $X, Y \in M$, we define $D(X, Y)$ as the infimum of $\delta > 0$ such that there are continuous maps $i : X \to Y$ and $j : Y \to X$ such that

$$d(i \circ j, I) \le \delta \quad \text{and} \quad d(j \circ i, I) \le \delta.$$

Example 1.9 The map D defined as above is normalized.

Proof If $i : X \to \{*\}$ and $j : \{*\} \to X$, then $d(i \circ j(*), *) = 0 < \mathrm{diam}(X)$ and $d(j \circ i(x), x) = d(j(*), x) \le \mathrm{diam}(X)$ for every $x \in X$. Thus, $D(X, Y) \le \mathrm{diam}(X)$.

Next, we show that

$$\inf_{x' \in X} \sup_{x \in X} d(x, x') \ge \frac{1}{3} \mathrm{diam}(X) \tag{1.1}$$

for all $X \in M$. Indeed, suppose that the inequality in (1.1) fails. Then there exists $x' \in X$ such that $d(x, x') < \frac{1}{3} \mathrm{diam}(X)$ for every $x \in X$. So, for every $x, x'' \in X$,

$$d(x, x'') \le d(x, x') + d(x'', x') < \frac{2}{3} \mathrm{diam}(X).$$

Since X is compact, we get a contradiction, proving (1.1).

Now suppose that $D(X, \{*\}) < \frac{1}{3} \mathrm{diam}(X)$. Then there are $\Delta < \frac{1}{3} \mathrm{diam}(X)$ and maps $i : X \to \{*\}$ and $j : \{*\} \to X$ such that $d(i \circ j, I) \le \Delta$ and $d(j \circ i, I) \le \Delta$. By taking $x' = j(*)$, we get $d(x', x) \le \Delta$ for every $x \in X$. Thus, $\sup_{x \in X} d(x', x) \le \Delta < \frac{1}{3} \mathrm{diam}(X)$, contradicting (1.1). This implies that $D(X, \{*\}) \ge \frac{1}{3} \mathrm{diam}(X)$. We conclude that $D(X, \{*\}) \le diam(X) \le 3D(X, Y)$ and so D is normalized (with $L = 3$). □

Now we prove that being normalized is *incompatible* with the properties *(P1)–(P3)*. More precisely, we have the following result.

Theorem 1.7 *A map* $D : M \times M \to [0, \infty[$ *satisfying (P1)–(P3) cannot be normalized.*

Proof We assume by contradiction that D satisfies (P1)-(P3) and is normalized. Take $X \in M$ with a positive diameter and $X' \in M$ with $X \equiv X'$. If $Y \in M$, then

$$D(X', Y) \overset{(P3)}{\le} K(D(X', X) + D(X, Y)) \overset{(P1)}{=} KD(X, Y).$$

Reversing the roles of X and X' using (P2), we get

$$K^{-1}D(X', Y) \le D(X, Y) \le KD(X', Y), \qquad \forall Y \in M. \tag{1.2}$$

Now take L from the normalization condition on D and $\lambda > L^2 K$. Define X' as X endowed with the metric $d'(x, x') = \lambda d(x, x')$ for $x, x' \in X$, where d is the metric of X. Then X' is a compact metric space with $\operatorname{diam}(X') = \lambda, \operatorname{diam}(X)$. Furthermore, the identity $i : X \to X'$, that is, $i(x) = x$ for $x \in X$, is a homeomorphism. Then $X \equiv X'$. Therefore, from the normalization inequality, we obtain

$$\lambda, \operatorname{diam}(X) = \operatorname{diam}(X') \le LD(X', \{*\}) \overset{(1.2)}{\le} KLD(X, \{*\}) \le KL^2\operatorname{diam}(X).$$

Dividing by $\operatorname{diam}(X) > 0$, we get $\lambda \le L^2 K$, a contradiction. This completes the proof.
□

Theorem 1.7 implies.

Example 1.10 The map D defined in Example *1.9* does not satisfy *(P3)*.

See also Exercise 1.27.

1.7 C^0-Gromov-Hausdorff Distance for Dynamical Systems

In this section, we introduce the C^0-Gromov-Hausdorff distance for certain dynamical systems. This includes maps, flows, and finitely generated group actions. When necessary, we will let d^X denote the distance on a metric space X.

Let $f : X \to X$ and $g : Y \to Y$ be continuous maps of compact metric spaces. We say that f and g are *isometrically conjugate* if there is an isometry $h : Y \to X$ such that $f \circ h = h \circ g$. In the following theorem, we characterize isometrically conjugate maps.

Theorem 1.8 *Two maps $f : X \to X$ and $g : Y \to Y$ of compact metric spaces are isometrically conjugate if and only if*

$$\inf\{\Delta > 0 : \exists \Delta\text{-isometry } i : X \to Y \text{ such that } d(g \circ i, i \circ f) \le \Delta\} = 0.$$

Proof Suppose that f and g are isometrically conjugate. Namely, there is an isometry $h : Y \to X$ such that $f \circ h = h \circ g$. Then for any $\Delta > 0$, $h : Y \to X$ and $h^{-1} : Y \to X$ are Δ-isometries satisfying $d(g \circ h^{-1}, h^{-1} \circ f) = d(h \circ g, f \circ h) = 0 < \Delta$. Since Δ is arbitrary, the infimum in the statement is zero.

Conversely, suppose that the infimum in the statement is zero. Then there are sequences of $\frac{1}{n}$-isometries $\{i_n : X \to Y\}_{n \in \mathbb{N}}$ and $\{j_n : Y \to X\}_{n \in \mathbb{N}}$ satisfying

$$d(g \circ i_n, i_n \circ f) < \frac{1}{n} \quad \text{and} \quad d(j_n \circ g, f \circ j_n) < \frac{1}{n}$$

for every $n \in \mathbb{N}$. Select a countable dense subset A of X; without loss of generality, we can assume that $f(A) \subset A$. By a diagonal argument, since Y is compact, we can choose a subsequence $\{i_{n_l}\}$ of $\{i_n\}_{n \in \mathbb{N}}$ such that $i_{n_l}(a)$ converges in Y for all $a \in A$. Denoting the limit of $i_{n_l}(a)$ by $i(a)$, we obtain a map $i : A \to Y$ satisfying

$$i(a) = \lim_{l \to \infty} i_{n_l}(a)$$

for all $a \in A$. But since each i_n is a $\frac{1}{n}$-isometry, we have that

$$d(a, a') - \frac{1}{n_l} < d(i_{n_l}(a), i_{n_l}(a')) < d(a, a') + \frac{1}{n_l}$$

for all $l \in \mathbb{N}$ and $a, a' \in A$. Letting $l \to \infty$, we obtain

$$d(i(a), i(a')) = d(a, a').$$

From this and the denseness of A, we obtain an extension of i to the whole X, still denoted by $i : X \to Y$, such that

$$d(i(x), i(x')) = d^X(x, x')$$

for all $x, x' \in X$. Since i_n is a $\frac{1}{n}$-isometry for all $n \in \mathbb{N}$, we see that i is onto. Therefore, i is an isometry.

Next, since

$$d(g(i_{n_l}(a)), i_{n_l}(f(a))) < \frac{1}{n_l}$$

for all $l \in \mathbb{N}$ and $a \in A$, $f(a) \in A$ for all $a \in A$, $i_{n_l}(a) \to i(a)$ as $l \to \infty$, and g is continuous, we obtain by letting $l \to \infty$ above that $d(g(i(a)), i(f(a)) = 0$ for every $a \in A$. Therefore, $g \circ i = i \circ f$ in A. Since A is dense in X, $g \circ i = i \circ f$ in X. Hence, f and g are isometrically conjugated. This completes the proof. □

Theorem 1.8 suggests the infimum in its statement can serve as a kind of measure of how far two given maps isometric can be. This suggests the following concept.

Definition 1.11 The C^0-*Gromov-Hausdorff distance* between two continuous maps $f : X \to X$ and $g : Y \to Y$ of compact metric spaces is defined by

$$d_{\mathrm{GH}^0}(f, g) = \inf\{\Delta > 0 : \text{there exist } \Delta\text{-isometries } i : X \to Y \text{ and } j : Y \to X$$

$$\text{such that } d(g \circ i, i \circ f) < \Delta \text{ and } d(j \circ g, f \circ j) < \Delta\}.$$

The following remark holds.

Remark 1.1 From the above definition, we obtain a notion of convergence for maps: If $\{Y_k\}_{k \in \mathbb{N}}$ is a sequence of metric spaces, we say that a sequence of maps $\{g_k : Y_k \to Y_k\}_{k \in \mathbb{N}}$ converges to $f : X \to X$ (momentarily denoted by $g_k \overset{\mathrm{GH}^0}{\to} f$) if $d_{\mathrm{GH}^0}(f, g_k) \to 0$ as $k \to \infty$. Now, $g_k \overset{\mathrm{GH}^0}{\to} f$ does not imply that g_k converges to f in the sense of Gromov [61]. In fact, of $f : X \to X$ is a map, then the constant sequence $\{g_k\}_{k \in \mathbb{N}}$ defined by $g_k = f$ for every $k \in \mathbb{N}$ always satisfies $g_k \overset{\mathrm{GH}^0}{\to} f$, but g_k converges to f in the Gromov sense only if f is continuous (cf. [61, p. 401]).

Next we prove some properties of the C^0-Gromov-Hausdorff distance. We will use the notation $f \approx g$ to indicate that two continuous maps $f : X \to X$ and $g : Y \to Y$ of compact metric spaces are isometrically conjugate.

Theorem 1.9 *The following properties hold for every pair of continuous maps $f : X \to X$ and $g : Y \to Y$ of compact metric spaces:*

(1) *If $X = Y$, then $d_{\mathrm{GH}^0}(f, g) \leq d(f, g)$.*
(2) *$d_{\mathrm{GH}}(X, Y) \leq d_{\mathrm{GH}^0}(f, g)$ and $d_{\mathrm{GH}}(X, Y) = d_{\mathrm{GH}^0}(\mathrm{Id}_X, \mathrm{Id}_Y)$, where Id_Z is the identity map of Z.*
(3) *$d_{\mathrm{GH}^0}(f, g) = 0$ if and only if $f \approx g$.*
(4) *$d_{\mathrm{GH}^0}(f, g) = d_{\mathrm{GH}^0}(g, f)$.*
(5) *For every continuous map $r : Z \to Z$ of a compact metric space, we have*

$$d_{\mathrm{GH}^0}(f, r) \leq 2(d_{\mathrm{GH}^0}(f, g) + d_{\mathrm{GH}^0}(g, r)).$$

(6) *$0 \leq d_{\mathrm{GH}^0}(f, g) < \infty$.*
(7) *If there is a sequence of isometries $\{g_n : Y_n \to Y_n\}_{n \in \mathbb{N}}$ such that $d_{\mathrm{GH}^0}(f, g_n) \to 0$ as $n \to \infty$, then f is also an isometry.*

Proof First, we prove Item (1). Suppose $X = Y$. Take $\varepsilon > 0$, $\Delta = d(f, g) + \varepsilon$, and $i = j = \mathrm{Id}_X$. Then i and j are $d(f, g)$-isometries satisfying $d(g \circ i, i \circ f) = d(j \circ g, f \circ j) = d(f, g) < \Delta$. Therefore, $d_{\mathrm{GH}^0}(f, g) \le \Delta = d(f, g) + \varepsilon$. Since ε is arbitrary, $d_{\mathrm{GH}^0}(f, g) \le d(f, g)$.

Let us prove Item (2). It follows easily from the definitions that $d_{\mathrm{GH}}(X, Y) \le d_{\mathrm{GH}^0}(f, g)$. On the other hand, fix $\varepsilon > 0$. Then there are $\Delta < d_{\mathrm{GH}}(X, Y) + \varepsilon$ and Δ-isometries $i : X \to Y$ and $j : Y \to X$. Clearly, $d(\mathrm{Id}_Y \circ i, i \circ \mathrm{Id}_X) = d(j \circ \mathrm{Id}_Y, \mathrm{Id}_X \circ j) = 0 < \Delta$, and so $d_{\mathrm{GH}^0}(\mathrm{Id}_X, \mathrm{Id}_Y) \le \Delta < d_{\mathrm{GH}}(X, Y) + \varepsilon$. Since ε is arbitrary, $d_{\mathrm{GH}^0}(\mathrm{Id}_X, \mathrm{Id}_Y) \le d_{\mathrm{GH}}(X, Y)$. Consequently, that $d_{\mathrm{GH}^0}(\mathrm{Id}_X, \mathrm{Id}_Y) = d_{\mathrm{GH}}(X, Y)$.

Item (3) is a direct consequence of Theorem 1.8 and Item (4) is a direct consequence of the definition.

Let us prove Item (5). Fix $\varepsilon > 0$. It follows from the definition that there are $0 < \Delta_1 < d_{\mathrm{GH}^0}(f, g) + \varepsilon$, $0 < \Delta_2 < d_{\mathrm{GH}^0}(g, r) + \varepsilon$, Δ_1-isometries $i : X \to Y$, $j : Y \to X$, and Δ_2-isometries $k : Y \to Z$, $l : Z \to Y$ such that

$$d(g \circ i, i \circ f) < \Delta_1, \qquad d(j \circ g, f \circ j) < \Delta_1,$$

$$d(r \circ k, k \circ g) < \Delta_2, \quad \text{and} \quad d(l \circ r, g \circ l) < \Delta_2.$$

By Proposition 1.1, $k \circ i : X \to Z$ and $l \circ j : Z \to X$ are $2(\Delta_1 + \Delta_2)$-isometries. Since

$$
\begin{aligned}
d(r \circ k \circ i, k \circ i \circ f) &\le d(r \circ k \circ i, k \circ g \circ i) + d(k \circ g \circ i, k \circ i \circ f) \\
&\le 2\Delta_2 + d(g \circ i, i \circ f) \\
&\le 2(\Delta_1 + \Delta_2),
\end{aligned}
$$

we get

$$d(r \circ k \circ i, k \circ i \circ f) \le 2(\Delta_1 + \Delta_2).$$

Likewise,

$$d(l \circ j \circ r, f \circ l \circ j) \le 2(\Delta_1 + \Delta_2),$$

proving that

$$d_{\mathrm{GH}^0}(f, r) \le 2(\Delta_1 + \Delta_2) < 2(d_{\mathrm{GH}^0}(f, g) + d_{\mathrm{GH}^0}(g, r) + 2\varepsilon).$$

Letting $\varepsilon \to 0$, we get Item (5).

To prove Item (6), we first observe that $d_{\mathrm{GH}^0}(f, g) \geq 0$ is obvious by definition. The finiteness of $d_{\mathrm{GH}^0}(f, g)$ is a direct consequence of the inequality

$$d_{\mathrm{GH}^0}(f, g) \leq \max\{d_{\mathrm{GH}}(X, Y), \mathrm{diam}(X), \mathrm{diam}(Y)\}.$$

(see Exercise 1.23).

Let us prove Item (7). Since $d_{\mathrm{GH}^0}(f, g_n) \to 0$, there are a sequence $\{\delta_n\}_{n \in \mathbb{N}}$ with $\delta_n \to 0$ and δ_n-isometries $i_n : X \to Y_n$ and $j_n : Y_n \to X$ such that

$$d(g_n \circ i_n, i_n \circ f) < \delta_n \quad \text{and} \quad d(j_n \circ g_n, f \circ j_n) < \delta_n$$

for every $n \in \mathbb{N}$. It follows that

$$\begin{aligned}
d^{Y_n}(i_n(f(x)), i_n(f(x'))) &\leq d^{Y_n}(g_n(i_n(x)), i_n(f(x))) \\
&\quad + d^{Y_n}(g_n(i_n(x)), g_n(i_n(x'))) \\
&\quad + d^{Y_n}(g_n(i_n(x')), i_n(f(x'))) \\
&< 3\delta_n + d^X(x, x')
\end{aligned}$$

for all $x, x' \in X$. Replacing here $d^X(f(x), f(x')) < \delta_n + d^{Y_n}(i_n(f(x)), i_n(f(x')))$, we obtain

$$d^X(f(x), f(x')) < 4\delta_n + d^X(x, x').$$

Letting $n \to \infty$, we get

$$d^X(f(x), f(x')) \leq d^X(x, x'). \tag{1.3}$$

On the other hand,

$$d_{\mathrm{H}}^X(f(j_n(Y_n)), X) \leq d_{\mathrm{H}}^X(f(j_n(Y_n)), j_n(g_n(Y_n))) + d_{\mathrm{H}}^X(j_n(g_n(Y_n)), X) \tag{1.4}$$

for all $n \in \mathbb{N}$. But $g_n(Y_n) = Y_n$, since g_n is an isometry. Hence,

$$d_{\mathrm{H}}^X(j_n(g_n(Y_n)), X) = d_{\mathrm{H}}^X(j_n(Y_n), X) \to 0 \quad \text{as} \quad n \to \infty,$$

since j_n is a δ_n-isometry with $\delta_n \to 0$. As $d(j_n \circ g_n, f \circ j_n) < \delta_n \to 0$, we obtain

$$d_{\mathrm{H}}^X(f(j_n(Y_n)), j_n(g_n(Y_n))) \to 0.$$

Letting $n \to \infty$ in (1.4), we see that $d_{\mathrm{H}}^X(f(j(Y_n)), X) \to 0$. This shows that f is onto and so an isometry by (1.3) and [26]. This completes the proof. □

One more property of the C^0-Gromov-Hausdorff distance between compact metric spaces is given below.

Lemma 1.5 *Let $X, Y, X', Y' \in M$ and $f : X \to X$, $g : Y \to Y$, $f' : X' \to X'$ and $g' : Y' \to Y'$ be continuous. If $f \approx f'$ and $g \approx g'$, then $d_{GH^0}(f, g) = d_{GH^0}(f', g')$.*

Proof By hypothesis, there is an isometry $B : Y \to Y'$ such that $g' \circ B = B \circ g$. Fix $\delta > 0$. Then there are $(d_{GH^0}(f, g) + \delta)$-isometries $i : X \to Y$ and $j : Y \to X$ such that $d(g \circ i, i \circ f) < d_{GH^0}(f, g) + \delta$ and $d(g \circ j, j \circ f) < d_{GH^0}(f, g) + \delta$. By Lemma 1.1, $B \circ i : X \to Y'$ and $j \circ B^{-1} : Y' \to X$ are $(d_{GH^0}(f, g) + \delta)$-isometries. Since

$$d(g' \circ B \circ i, B \circ i \circ f) = d(B \circ g \circ i, B \circ i \circ f) = d(g \circ i, i \circ f)$$
$$< d_{GH^0}(f, g) + \delta$$

and

$$d(j \circ B^{-1} \circ g', f \circ j \circ B^{-1}) = d(j \circ g \circ B^{-1}, f \circ j \circ B^{-1}) = d(j \circ g, f \circ j)$$
$$< d_{GH^0}(f, g) + \delta,$$

we get $d_{GH^0}(f, g') \leq d_{GH^0}(f, g) + \delta$. Since δ is arbitrary, $d_{GH^0}(f, g') \leq d_{GH^0}(f, g)$. Reversing the roles of g and g', we get $d_{GH^0}(f, g) \leq d_{GH^0}(f, g')$. So, $d_{GH^0}(f, g) = d_{GH^0}(f, g')$, which implies that $d_{GH^0}(f, g) = d_{GH^0}(f, g') = d_{GH^0}(f', g')$.   □

A lemma closely related to the one below was stated without proof as Lemma 2.5 in [29, p.523].

Lemma 1.6 *For any $\delta > 0$ and any δ-isometry $j : Y \to X$ of compact metric spaces, there is a 3δ-isometry $i : X \to Y$ such that*

$$d(j \circ i, I) \leq \delta \quad and \quad d(i \circ j, I) \leq 2\delta.$$

Proof Since j is a δ-isometry, for each $x \in X$ we can select $i(x) \in Y$ such that $d(j(i(x)), x) \leq \delta$. This yields a map $i : X \to Y$ such that

$$d(j \circ i, I) \leq \delta.$$

Next, we prove that i is a 3δ-isometry. On the one hand, since

$$d(i(x), i(x')) < \delta + d(j(i(x)), j(i(x')))$$
$$\leq \delta + d(j(i(x)), x) + d(j(i(x')), x') + d(x, x')$$
$$\leq 3\delta + d(x, x')$$

and

$$d(x, x') < d(j(i(x)), x) + d(j(i(x)), j(i(x'))) + d(j(i(x')), x')$$
$$< 3\delta + d(i(x), i(x'))$$

for every $x, x' \in X$, we have

$$\sup_{x,x'\in X} |d(i(x), i(x')) - d(x, x')| \leq 3\delta.$$

On the other hand, if $y \in Y$, then by taking $x = j(y)$ we have $d(i(x), y) < \delta + d(j(ix)), x) \leq 2\delta$, implying that

$$d_H(i(X), Y) \leq 2\delta.$$

Therefore, i is a 3δ-isometry. Finally,

$$d(i(j(y)), y) < \delta + d(j \circ i(j(y)), j(y)) \leq \delta + \delta = 2\delta$$

for all $y \in Y$. So, $d(i \circ j, I) < 2\delta$, as needed. $\qquad\square$

A *flow* on a metric space X is a continuous map $\phi : X \times \mathbb{R} \to X$ such that $\phi(x, 0) = x$ and $\phi(\phi(x, s), t) = \phi(x, s + t)$ for any $x \in X$ and $s, t \in \mathbb{R}$. The set of flows (X, ϕ) on compact metric spaces will be denoted by $\mathcal{CDS}$.

For any (X, ϕ) and (Y, ψ) in $\mathcal{CDS}$, we say that ϕ and ψ are isometric if there exists an isometry $i : X \to Y$ such that $i(\phi(x, t)) = \psi(i(x), t)$ for all $x \in X$ and $t \in [0, 1]$. For any $\varepsilon > 0$ and $X \in \mathcal{M}$, we denote by $\text{Rep}_X(\varepsilon)$ the collection of all continuous maps $\alpha : X \times \mathbb{R} \to \mathbb{R}$ such that for each fixed $x \in X$, $\alpha(x, \cdot) : \mathbb{R} \to \mathbb{R}$ is a homeomorphism satisfying $\alpha(x, 0) = 0$ for all $x \in X$ and $|\alpha(x, t) - t| < \varepsilon$ for all $t \in \mathbb{R}$; such as α is called a reparanetrifation.

In this section, we introduce a topology on $\mathcal{CDS}$ based on the following definition.

Definition 1.12 We define the *Gromov-Hausdorff distance* $D_{\text{GH}^0}(\phi, \psi)$ between $(X, \phi), (Y, \psi) \in \mathcal{CDS}$ as the infimum of the numbers $\varepsilon > 0$ for which there exist two ε-isometries $i : X \to Y$ and $j : Y \to X$, and $\alpha \in \text{Rep}_X(\varepsilon)$ and $\beta \in \text{Rep}_Y(\varepsilon)$ such that for any $x \in X$, $y \in Y$, and $t \in [0, 1]$,

$$d(i(\phi(x, \alpha(x, t))), \psi(i(x), t)) < \varepsilon \quad \text{and} \quad d(j(\psi(y, \beta(y, t))), \phi(j(y), t)) < \varepsilon.$$

This definition differs from the definition of the C^0-Gromov-Hausdorff distance in [21], where continuous ε-isometries (but not representations) were considered. This results in a distance between flows that is greater than the one introduced above.

Now we will prove several basic properties extending Theorem 1.9.

Proposition 1.2 *For any* $(X, \phi), (Y, \psi), (Z, \xi) \in \mathcal{CDS}$, *the following properties hold:*

(1) $D_{\mathrm{GH}^0}(\phi, \psi) \geq 0$; *equality holds if and only if ϕ and ψ are isometric.*
(2) $D_{\mathrm{GH}^0}(\phi, \psi) = D_{\mathrm{GH}^0}(\psi, \phi)$.
(3) $D_{\mathrm{GH}^0}(\phi, \psi) \leq 2(D_{\mathrm{GH}^0}(\phi, \xi) + D_{\mathrm{GH}^0}(\xi, \psi))$.

Proof Proof of (1). If ϕ and ψ are isometric, then clearly $D_{\mathrm{GH}^0}(\phi, \psi) = 0$.

Suppose $D_{\mathrm{GH}^0}(\phi, \psi) = 0$. Then for any $n \in \mathbb{N}$, there are $\frac{1}{n}$-isometries $i_n : X \to Y$ and representations $j_n : Y \to X$, and $\alpha_n \in \mathrm{Rep}_X(1/n)$ and $\beta_n \in \mathrm{Rep}_Y(1/n)$ such that for any $x \in X$, $y \in Y$, and $t \in [0, 1]$,

$$d(i_n(\phi(x, \alpha_n(x, t))), \psi(i_n(x), t)) < \frac{1}{n}, d(j_n(\psi(y, \beta_n(y, t))), \phi(j_n(y), t)) < \frac{1}{n}.$$

For a countable dense subset A_0 of X, let $A = \phi(A_0, \mathbb{Q})$. Then A is also a countable dense subset of X such that $\phi(A, t) \subset A$ for any $t \in \mathbb{Q}$. By the Cantor diagonal argument, we may assume that there is a subsequence $\{i_{n_k}\}_{k \in \mathbb{N}}$ of $\{i_n\}_{n \in \mathbb{N}}$ such that for any $a \in A$, $i_{n_k}(a)$ converges to a point, say $i(a) \in Y$. Then we can define a map $i : A \to Y$ by $i(a) := \lim_{k \to \infty} i_{n_k}(a)$ for any $a \in A$. Since

$$d(i(a), i(a')) \leq \lim_{k \to \infty} \left(d(a, a') + \frac{1}{n_k} \right) = d(a, a'),$$

we see that $d(i(a), i(a')) = d(a, a')$ for any $a, a' \in A$, and so i is uniformly continuous. Hence, we can extend it to a uniformly continuous map on X satisfying $d(i(x), i(x')) = d(x, x')$ for any $x, x' \in X$. On the other hand, we see that i is surjective. Indeed, for any $y \in Y$, take a sequence $\{a_{n_k}\}_{k \in \mathbb{N}}$ in A such that $i_{n_k}(a_{n_k})$ converges to y. We may assume that a_{n_k} converges to $x \in X$. Since

$$d(i(x), y) \leq d(i(x), i(a_{n_k})) + d(i(a_{n_k}), i_{n_k}(a_{n_k})) + d(i_{n_k}(a_{n_k}), y),$$

we see that $d(i(x), y) = 0$, and so i is an isometry.

For any $a \in A$ and $t \in \mathbb{Q}$, we have

$$\begin{aligned}
d(\psi(i(a), t), i(\phi(a, t))) \leq\ & d(\psi(i(a), t), \psi(i_{n_k}(a), t)) \\
& + d(\psi(i_{n_k}(a), t), i_{n_k}(\phi(a, \alpha_{n_k}(a, t)))) \\
& + d(i_{n_k}(\phi(a, \alpha_{n_k}(a, t))), i_{n_k}(\phi(a, t))) \\
& + d(i_{n_k}(\phi(a, t)), i(\phi(a, t))).
\end{aligned}$$

Letting $k \to \infty$, we obtain $\psi(i(a), t) = i(\phi(a, t))$. So, $\psi(i(x), t) = i(\phi(x, t))$ for all $x \in X$ and $t \in \mathbb{R}$.

We can define $j : Y \to X$ such that $j(\psi(y, t)) = \phi(j(y), t)$ for all $y \in Y$ and $t \in \mathbb{R}$. This shows that ϕ and ψ are isometric.

Proof of (2). The symmetry is clear by the definition.

Proof of (3). For any fixed $\varepsilon > 0$, we let $\delta_1 = D_{GH^0}(\phi, \xi) + \varepsilon$. Then there are δ_1-isometries $i_1 : X \to Z$ and $j_1 : Z \to X$, and $\alpha_1 \in \text{Rep}_X(\delta_1)$ and $\beta_1 \in \text{Rep}_Z(\delta_1)$ such that for any $x \in X, z \in Z$, and $t \in [0, 1]$,

$$d(i_1(\phi(x, \alpha_1(x, t))), \xi(i_1(x), t)) < \delta_1$$

and

$$d(j_1(\xi(z, \beta_1(z, t))), \phi(j_1(z), t)) < \delta_1.$$

Let $\delta_2 = D_{GH^0}(\xi, \psi) + \varepsilon$. Then there are δ_2-isometries $i_2 : Z \to Y$ and $j_2 : Y \to Z$, and $\alpha_2 \in \text{Rep}_Z(\delta_2)$ and $\beta_2 \in \text{Rep}_Y(\delta_2)$ such that for any $y \in Y, z \in Z$, and $t \in [0, 1]$,

$$d(i_2(\xi(z, \alpha_2(z, t))), \psi(i_2(z), t)) < \delta_2$$

and

$$d(j_2(\psi(y, \beta(y, t))), \xi(j_2(y), t)) < \delta_2.$$

We note that $i_2 \circ i_1$ and $j_1 \circ j_2$ are $2(\delta_1 + \delta_2)$-isometries. Now define $\alpha_1 \circ \alpha_2 \in \text{Rep}_X(\delta_1 + \delta_2)$ by

$$\alpha_1 \circ \alpha_2(x, t) := \alpha_1(x, \alpha_2(i_1(x), t))$$

for all $(x, t) \in X \times \mathbb{R}$. For any $x \in X$ and $t \in [0, 1]$, we have

$$d(\psi(i_2 \circ i_1(x), t), i_2 \circ i_1(\phi(x, \alpha_1 \circ \alpha_2(x, t))))$$
$$\leq d(\psi(i_2 \circ i_1(x), t), i_2(\xi(i_1(x), \alpha_2(i_1(x), t))))$$
$$+ d(i_2(\xi(i_1(x), \alpha_2(i_1(x), t))), i_2 \circ i_1(\phi(x, \alpha_1 \circ \alpha_2(x, t))))$$
$$\leq \delta_2 + \delta_2 + d(\xi(i_1(x), \alpha_2(i_1(x), t)), i_1(\phi(x, \alpha_1 \circ \alpha_2(x, t))))$$
$$< 2(\delta_1 + \delta_2).$$

Define $\beta_2 \circ \beta_1 \in \text{Rep}_Y(\delta_1 + \delta_2)$ by

$$\beta_2 \circ \beta_1(y, t) := \beta_2(y, \beta_1(j_2(y), t))$$

for all $(y, t) \in Y \times \mathbb{R}$. Then

$$d(\phi(j_1 \circ j_2(y), t), j_1 \circ j_2(\psi(y, \beta_2 \circ \beta_1(y, t)))) < 2(\delta_1 + \delta_2),$$

where $y \in Y$ and $t \in [0, 1]$. Consequently,

$$D_{\mathrm{GH}^0}(\phi, \psi) < 2(\delta_1 + \delta_2) = 2(D_{\mathrm{GH}^0}(\phi, \xi) + D_{\mathrm{GH}^0}(\xi, \psi)) + 4\varepsilon.$$

Since ε is arbitrary, we get $D_{\mathrm{GH}^0}(\phi, \psi) \leq 2(D_{\mathrm{GH}^0}(\phi, \xi) + D_{\mathrm{GH}^0}(\xi, \psi))$.   □

Proposition 1.2, slows that the distance D_{GH^0} is a quasi-metric on $\mathcal{CDS}$. If we use the main result of the paper by Paluszynski and Stempak [59], then we can get a metric $d_{1/2}$ on $\mathcal{CDS}$ such that

$$d_{1/2}(\phi, \psi) \leq \left(D_{\mathrm{GH}^0}(\phi, \psi)\right)^{1/2} \leq 4 d_{1/2}(\phi, \psi)$$

for any $\phi, \psi \in \mathcal{CDS}$. The topology on $\mathcal{CDS}$ induced by the metric $d_{1/2}$ will be called the topology of Gromov-Hausdorff convergence, which we denote by $\mathcal{T}_{GH}$.

Let G be a group with neutral element e. We say that $A \subset G$ *generates* G if every $g \in G$ is the product of finitely many elements of A. Hereafter, generating sets A will be assumed to be *symmetric*, that is, $a^{-1} \in A$ for every $a \in A$. We say that G is *finitely generated* if it has a finite generating set.

A G-*action* on X is a continuous map $T : G \times X \to X$ such that $T(e, x) = x$ and $T(g, T(h, x)) = T(gh, x)$ for all $x \in X$ and $g, h \in G$. By writing $T_g(x) = T(g, x)$, we identify a G-action T with a parametrized family of homeomorphisms $\{T_g : X \to X\}_{g \in G}$ such that $T_e = \mathrm{Id}_X$ (the identity of X) and $T_{gh} = T_g \circ T_h$ for every $g, h \in G$. Denote by $\mathrm{Act}(G, X)$ the set of G-actions T on X.

Definition 1.13 Given compact metric spaces X and Y, a nonempty subset $A \subset G$, $T \in \mathrm{Act}(G, X)$, and $S \in \mathrm{Act}(G, Y)$, we define $d_{\mathrm{GH}^0, A}(T, S)$ as the infimum of the numbers $\Delta > 0$ for which there are Δ-isometries $i : X \to Y$ and $j : Y \to X$ such that

$$\sup_{a \in A, x \in X} d^Y(S_a \circ i(x), i \circ T_a(x)) < \Delta$$

and

$$\sup_{a \in A, y \in Y} d^X(j \circ S_a(y), T_a \circ j(y)) < \Delta.$$

This definition is a slight variation of that of the equivariant Hausdorff distance for compact metric spaces equipped with isometric actions introduced by Fukaya [29]. The difference is that, unlike the equivariant distance, the above distance does not consider the whole group G, but only a finite subset (usually a generator) A instead. As a result, we obtain a topology that is stronger than the one induced by the equivariant distance. We will prove in Lemma 1.9 that the topologies furnished by different generators are equivalent. By taking $G = \mathbb{Z}$ and $A = \{1, -1\}$, we recover the Gromov-Hausdorff distance for homeomorphisms defined before. Fukaya's distance for not necessarily equivariant group actions was recently considered by Chung [22].

Some of the properties of the Gromov-Hausdorff distance for homeomorphisms can be extended to the context of group actions. It suffices to know only the ones proved below.

Lemma 1.7 *Let X, Y, Z be compact metric spaces, G be a finitely generated group, and $A \subset G$.*

(1) *If $R \in \mathrm{Act}(G, Z)$, then*

$$d_{\mathrm{GH}^0, A}(T, S) \le 2(d_{\mathrm{GH}^0, A}(T, R) + d_{\mathrm{GH}^0, A}(R, S)).$$

(2) *If A is a generating set, then $d_{\mathrm{GH}^0, A}(T, S) = 0$ if and only if T and S are isometrically conjugated.*

Proof First, we prove Item (1). Fix $\varepsilon > 0$. Take Δ_1 and Δ_2 such that

$$0 < \Delta_1 < d_{\mathrm{GH}^0, A}(T, R) + \varepsilon \quad and \quad 0 < \Delta_2 < d_{\mathrm{GH}^0, A}(R, S) + \varepsilon.$$

Then there are Δ_1-isometries $i : X \to Z$ and $j : Z \to X$, and Δ_2-isometries $k : Y \to Z$ and $l : Z \to Y$, such that for every $a \in A$,

$$d_{C^0}^Z(i \circ T_a, R_a \circ i) < \Delta_1,$$

$$d_{C^0}^X(j \circ R_a, T_a \circ j) < \Delta_1,$$

$$d_{C^0}^Z(R_a \circ k, k \circ S_a) < \Delta_2, \text{ and}$$

$$d_{C^0}^Y(l \circ R_a, S_a \circ l) < \Delta_2.$$

It follows that

$$d_{\mathrm{H}}(l \circ i(X), Y) < \Delta_1 + 2\Delta_2.$$

Indeed, since l is a Δ_2-isometry, if $y \in Y$, then there is $z \in Z$ such that $d(l(z), y) < \Delta_2$. Since i is a Δ_1-isometry, there is $x \in X$ such that $d(i(x), z) < \Delta_1$. Hence,

$$d(l(i(x)), l(z)) < \Delta_2 + d(i(x), z) < \Delta_2 + \Delta_1,$$

and consequently

$$d(l(i(x)), y) < d(l(i(x)), l(z)) + d(l(z), y)$$
$$< \Delta_2 + \Delta_1 + \Delta_2 = \Delta_1 + 2\Delta_2,$$

proving the assertion.

On the other hand,

$$|d^Y(l \circ i(x), l \circ i(x')) - d^X(x, x')| = |d^Y(l \circ i(x), l \circ i(x')) - d^Z(i(x), i(x'))$$
$$+ d^Z(i(x), i(x')) - d^X(x, x')|$$
$$\leq |d^Y(l \circ i(x), l \circ i(x')) - d^Z(i(x), i(x'))|$$
$$+ |d^Z(i(x), i(x')) - d^X(x, x')|$$
$$< \Delta_2 + \Delta_1.$$

Therefore, $l \circ i : X \to Y$ is a $(\Delta_1 + 2\Delta_2)$-isometry and also a $2(\Delta_1 + \Delta_2)$-isometry. Likewise, $j \circ k : Y \to X$ is a $2(\Delta_1 + \Delta_2)$-isometry.

Now, for all $x \in X$, we have

$$d^Y(S_a \circ l \circ i(x), l \circ i \circ T_a(x)) \leq d^Y(S_a \circ l \circ i(x), l \circ R_a \circ i(x))$$
$$+ d^Y(l \circ R_a \circ i(x), l \circ i \circ T_a(x))$$
$$< \Delta_2 + \Delta_1 + \Delta_2$$
$$< 2(\Delta_1 + \Delta_2).$$

Hence, $d^Y_{C^0}(S_a \circ l \circ i, l \circ i \circ T_a) \leq 2(\Delta_1 + \Delta_2)$. Likewise,

$$d^X_{C^0}(j \circ k \circ S_a, T_a \circ j \circ k) \leq 2(\Delta_1 + \Delta_2).$$

So,

$$d_{GH^0, A}(T, S) \leq 2(\Delta_1 + \Delta_2) < 2(d_{GH^0, A}(T, R) + \varepsilon + d_{GH^0, A}(R, S) + \varepsilon).$$

Since ε is arbitrary,

$$d_{GH^0, A}(T, S) \leq 2(d_{GH^0, A}(T, R) + d_{GH^0, A}(R, S)),$$

proving Item (1).

Next we prove Item (2). Assume that A is a finite generating set of G.

First, we prove the sufficiency. Suppose T and S are isometric. Then for any $\Delta > 0$, $h : Y \to X$ and $h^{-1} : X \to Y$ are Δ-isometries satisfying

$$d_{C^0}^Y(S_a \circ h^{-1}, h^{-1} \circ T_a) = d_{C^0}^X(T_a \circ h, h \circ S_a) = 0 < \Delta.$$

Therefore, $d_{\mathrm{GH}^0,A}(T, S) \leq \Delta$, and since Δ is arbitrary, $d_{\mathrm{GH}^0,A}(T, S) = 0$.

Next, we prove the necessity. Since $d_{\mathrm{GH}^0,A}(T, S) = 0$, there are sequences of $\frac{1}{n}$-isometries $\{i_n : X \to Y\}_{n \in \mathbb{N}}$ and $\{j_n : Y \to X\}_{n \in \mathbb{N}}$ such that

$$\max \left\{ \sup_{a \in A, x \in X} d^Y(S_a \circ i_n(x), i_n \circ T_a(x)), \sup_{a \in A, y \in Y} d^X(j_n \circ S_a(y), T_a \circ j_n(y)) \right\} < \frac{1}{n}$$

for all $n \in \mathbb{N}$. Since X is compact, we can select a countable dense subset B of X and assume $T_a(B) \subset B$ for every $a \in A$. By a diagonal argument, since Y is compact, we can choose a subsequence $\{i_{n_l}\}_{l \in \mathbb{N}}$ of $\{i_n\}_{n \in \mathbb{N}}$ such that $i_{n_l}(b)$ converges in Y for all $b \in B$.

Let $i(b)$ denote the limit of $i_{n_l}(b)$. Then we obtain a map $i : B \to Y$ rule that

$$i(b) = \lim_{l \to \infty} i_{n_l}(b) \text{ for every } b \in B.$$

Since i_{n_l} is a $\frac{1}{n_l}$-isometry, we have

$$|d^Y(i_{n_l}(x), i_{n_l}(x')) - d^X(x, x')| < \frac{1}{n_l}.$$

Then

$$d^X(x, x') - \frac{1}{n_l} < d^Y(i_{n_l}(x), i_{n_l}(x')) < d^X(x, x') + \frac{1}{n_l}$$

for all $l \in \mathbb{N}_+$ and $x, x' \in B$. Letting $l \to \infty$, we obtain

$$d^Y(i(x), i(x')) = d^X(x, x').$$

It follows that $i : B \to Y$ is an isometric immersion and thus uniformly continuous. Since B is dense in X, we obtain an extension of i to the whole X, still denoted by $i : X \to Y$, such that

$$d^Y(i(x), i(x')) = d^X(x, x')$$

for all $x, x' \in X$. Since i_n is a $\frac{1}{n}$-isometry for all n, we have

$$d_H(i(X), Y) = \max\left\{ \sup_{y\in i(X)} d(y, Y), \sup_{y\in Y} d(i(X), y) \right\} = 0$$

as $n \to \infty$. Consequently, i is surjective and so an isometry.

Next, since $T_a(b) \in B$ for $b \in B$, we have

$$d_{C^0}^Y(S_a \circ i_{n_l}(b), i_{n_l} \circ T_a(b)) < \frac{1}{n_l}$$

for all $l \in \mathbb{N}_+$ and $b \in B$. Since $i_{n_l}(b) \to i(b)$ as $l \to \infty$,

$$d_{C^0}^Y(S_a \circ i(b), i \circ T_a(b)) = 0.$$

Therefore, $S_a \circ i = i \circ T_a$ in B. Since B is dense in X, $S_a \circ i = i \circ T_a$ in X for all $a \in A$. Since A generates G, $S_g \circ i = i \circ T_g$ for all $g \in G$. Hence, T and S are isometrically conjugate. This completes the proof. $\square$

Our next task is to compare $d_{\mathrm{GH}^0,A}(T, S)$ and $d_{\mathrm{GH}^0,B}(T, S)$ for different generating sets A and B of G. This requires some preliminary definitions and facts.

Let X be a compact metric space and G be a finitely generated group. Given $T \in \mathrm{Act}(G, X)$ and $B \subset G$, we define the variation map $V : \mathbb{R}^+ \to \mathbb{R}^+$ by

$$V(\delta) = \sup_{b\in B}\{d^X(T_b(x), T_b(x')) : x, x' \in X, d^X(x, x') \le \delta\},$$

for $\delta \ge 0$. If necessary, we write $V_{T,B}$ instead of V to emphasize dependence on T and B. This map has two nice properties:

- V is increasing.
- If B is finite, then

$$\lim_{\delta \to 0} V(\delta) = 0. \tag{1.5}$$

The first is obvious and the second follows from the uniform continuity of T_b for all $b \in B$.

Let $B \subset G$. Define $B_0 = \emptyset$ and for $m \in \mathbb{N}$, let B_m be the set of all $g \in G$ that can be expressed as a product of at most m elements of B. Given a compact metric space Y, a map $j : Y \to X$, and an action $S \in \mathrm{Act}(G, Y)$, we define

$$\Delta_m(T, S, j, B) = \sup_{b_m\in B_m, y\in Y} d^X(j \circ S_{b_m}(y), T_{b_m} \circ j(y)) \tag{1.6}$$

for $m \in \mathbb{N}$. For simplicity of notation, we will write $\Delta_m = \Delta_m(T, S, j, B)$.

Lemma 1.8 *The following inequality holds for $m \in \mathbb{N}$:*

$$\Delta_m \leq \Delta_1 + V(\Delta_1 + V(\Delta_1 + \overset{(m)}{\cdots} + V(\Delta_1 + V(\Delta_1))\overset{(m)}{\cdots})).$$

Proof If $b_{m+1} \in B_{m+1}$, then $b_{m+1} = bb_m$ for some $b \in B_1$ and $b_m \in B_m$. Therefore,

$$d^X(j \circ S_{b_{m+1}}(y), T_{b_{m+1}} \circ j(y)) \leq d^X(j \circ S_b(S_{b_m}(y)), T_b \circ j(S_{b_m}(y)))$$

$$+ d^X(T_b(j \circ S_{b_m}(y)), T_b(T_{b_m} \circ j(y)))$$

$$< \Delta_1(T, S, j, B) + \sup\{d^X(T_b(x), T_b(x')) :$$

$$d^X(x, x') \leq \Delta_m, b \in B\}$$

$$= \Delta_1 + V(\Delta_m)$$

for all $y \in Y$, proving that $\Delta_{m+1} \leq \Delta_1 + V(\Delta_m)$ for all $m \in \mathbb{N}$. From this and a recursive argument, we get the result. □

The above lemma is used to prove the following.

Lemma 1.9 *Let X be a compact metric space, G a finitely generated group, and $T \in$ Act(G, X). Then for all finite generating set A and B of G and $\delta' > 0$, there is $\delta > 0$ such that if Y is a compact metric space and $S \in$ Act(G, Y) satisfies $d_{\mathrm{GH}^0, B}(T, S) < \delta$, then $d_{\mathrm{GH}^0, A}(T, S) < \delta'$.*

Proof Fix $\delta' > 0$. Since B is a finite generating set and A is finite, we can also fix $m \in \mathbb{N}$ such that $A \subset B_m$ (take, for instance, $m = \max_{a \in A} l_B(a)$, where l_B is the word length metric on G induced by B). Since m is fixed, (1.5) implies that there is a $\delta > 0$ satisfying

$$\max\{5\delta + V_{T,A}(2\delta), \delta + V_{T,B}(\delta + V_{T,B}(\delta + \overset{(m)}{\cdots} + V_{T,B}(\delta + V_{T,B}(\delta))\overset{(m)}{\cdots}))\} < \delta'.$$

Now, let Y be a compact metric space and take $S \in$ Act(G, Y) such that

$$d_{\mathrm{GH}^0, B}(T, S) < \delta.$$

It follows from the definitions that there is a δ-isometry $j : Y \to X$ such that

$$\Delta_1(T, S, j, B) < \delta.$$

Together with Lemma 1.8 and the choice of δ, we have

$$\Delta_m \leq \Delta_1 + V(\Delta_1 + V(\Delta_1 + \overset{(m)}{\cdots} + V(\Delta_1 + V(\Delta_1))\overset{(m)}{\cdots)})$$

$$< \delta + V_{T,B}(\delta + V_{T,B}(\delta + \overset{(m)}{\cdots} + V_{T,B}(\delta + V_{T,B}(\delta))\overset{(m)}{\cdots)})$$

$$< \delta'.$$

But $\delta < 5\delta < \delta'$. So, j is a δ'-isometry. Also, $A \subset B_m$; thus,

$$\sup_{a \in A, y \in Y} d^X(j \circ S_a(y), T_a \circ j(y)) \leq \sup_{b_m \in B_m, y \in Y} d^X(j \circ S_{b_m}(y), T_{b_m} \circ j(y)) = \Delta_m < \delta',$$

proving that

$$\sup_{a \in A, y \in Y} d^X(j \circ S_a(y), T_a \circ j(y)) < \delta'. \tag{1.7}$$

On the other hand, j is a δ-isometry. Hence, by Lemma 1.6, there is a 5δ-isometry $i : X \to Y$ such that

$$d^X(j \circ i(x), x) < 2\delta$$

for all $x \in X$. The choice of δ implies $5\delta < \delta'$; hence, i is also a δ'-isometry. Since

$$d^Y(S_a \circ i(x), i \circ T_a(x)) < \delta + d^X(j \circ S_a \circ i(x), j \circ i \circ T_a(x))$$

$$\leq \delta + d^X((j \circ S_a)(i(x)), (T_a \circ j)(i(x)))$$

$$+ d^X(T_a \circ j \circ i(x), j \circ i \circ T_a(x))$$

$$< 2\delta + d^X(T_a(j \circ i(x)), T_a(x)) + d^X(T_a(x), j \circ i(T_a(x)))$$

$$< 4\delta + d^X(T_a(j \circ i(x)), T_a(x))$$

for $x \in X$ and $a \in A$, we get

$$\sup_{a \in A, x \in X} d^Y(S_a \circ i(x), i \circ T_a(x)) < 4\delta + V_{T,A}(2\delta),$$

which in view of the choice of δ implies that

$$\sup_{a \in A, x \in X} d^Y(S_a \circ i(x), i \circ T_a(x)) < \delta'.$$

Then (1.7) implies $d_{\mathrm{GH}^0,A}(T, S) < \delta'$, proving the result. $\qquad\qquad\square$

To prove Theorem 2.7, we need the following lemma from [23].

Lemma 1.10 *For every compact metric space X, every countable group G, every expansive $T \in \text{Act}(G, X)$ with expansivity constant η, and every $\varepsilon > 0$, there exists a finite subset $F \subset G$ such that*

$$x, y \in X \text{ and } \sup_{g \in F} d(T_g x, T_g y) \leq \eta \quad \text{imply} \quad d(x, y) < \varepsilon.$$

Proof By contradiction, assume that there is $\varepsilon > 0$ such that for any non-empty finite set $F \subset G$, there are points $x_F, y_F \in X$ such that

$$\sup_{g \in F} d(T_g x, T_g y) \leq \eta \quad \text{but} \quad d(x_F, y_F) \geq \varepsilon.$$

Since G is countable, there is a sequence of non-empty finite subsets F_n of G such that

$$F_1 \subset F_2 \subset \cdots \quad \text{and} \quad G = \bigcup_{n \in \mathbb{N}} F_n.$$

Then, there are sequences $\{x_n\}_{n \in \mathbb{N}}$ and $\{y_n\}_{n \in \mathbb{N}}$ of F_n such that

$$\sup_{g \in F_n} d(T_g(x_n), T_g(y_n)) \leq \eta \quad \text{and} \quad d(x_n, y_x) \geq \varepsilon,$$

for $n \in \mathbb{N}$. By compactness, we can assume that $x_n \to x$ and $y_n \to y$ for some $x, y \in X$. It follows that $d(x, y) \geq \varepsilon$ (hence $x \neq y$), but $d(T_g(x), T_g(y)) \leq \eta$ for all $g \in G$. Since η is an expansivity constant, $x = y$, a contradiction. This completes the proof. $\square$

1.8 Gromov-Hausdorff Space of Continuous Maps

The *Gromov-Hausdorff space* is the set of compact metric spaces up to isometry $\mathcal{M}$ equipped with the Gromov-Hausdorff metric. It is known that this space is Polish, strictly intrinsic, and not boundedly compact [38, 39]. Also, generic elements in this space are totally discontinuous, totally anisometric, homeomorphic to the Cantor set, have no collinear triples of different points, and cannot be embedded into any Hilbert space [68].

Now consider the set of continuous maps of compact metric spaces up to isometric conjugacy. We will equip this set with the C^0-Gromov-Hausdorff distance defined in Sect. 1.5. The resulting space will be referred to as the *Gromov-Hausdorff space of continuous maps*. It is natural to ask whether the aforementioned properties of the Gromov-Hausdorff space also hold in the latter space. Some facts are known; for instance, the transitive elements in this space can be approximated by periodic orbits [24, 76].

More recently, it was proved that a generic element of the Gromov-Hausdorff space of continuous maps is topologically conjugated to a special homeomorphism defined of the Cantor set [10]. This extends part of the aforementioned results in [68] from the Gromov-Hausdorff space to the Gromov-Hausdorff space of continuous maps.

In this section, we will analyze the precompactness of subsets of the Gromov-Hausdorff space of continuous maps. More precisely, we prove that such a subset is precompact if and only if it is equicontinuous in a certain sense. Afterwards, we explain the characterization of continuous maps that can be approximated by periodic orbits. Our approach is new and even improves the one in [76]. Let us state these results in a precise way.

Consider a compact metric space X. Whenever necessary, we will denote the metric d of X by d^X. Given $A \subset X$ and $\varepsilon \geq 0$, we use the notation $A^\varepsilon = \{x \in X : \text{there exists } a \in A \text{ such that } d(a, x) < \varepsilon\}$.

Recall that $\mathcal{M}$ denotes the Gromov-Hausdorff space. Define

$$C = \{f : X \to X : f \text{ is continuous and } X \text{ is compact}\}.$$

We note that the isometric relation $f \approx g$ is an equivalence relation in C. We denote by $\mathcal{C}$ the corresponding set of equivalence classes. We call $\mathcal{C}$ the *Gromov-Hausdorff space of continuous maps*. Denoting by $[f] \in \mathcal{C}$ the equivalence class of $f \in C$, we can define by Lemma 1.5 the following quasi-metric in $\mathcal{C}$ for $[f], [g] \in \mathcal{C}$:

$$d_{\text{GH}^0}([f], [g]) = d_{\text{GH}^0}(f, g).$$

In what follows, we will write f instead of $[f]$, understanding that f is an equivalence class rather than a single map.

Now denote by $D(f) = X$ the map assigning the domain X to a map $f : X \to X$. Since isometrically conjugated homeomorphisms have isometric domains, there is an induced map $D : \mathcal{C} \to \mathcal{M}$. It follows easily from the definitions that

$$D_{\text{GH}^0}(D(f), D(g)) \leq d_{\text{GH}^0}(f, g), \qquad \forall f, g \in \mathcal{C}. \tag{1.8}$$

In particular, D is Lipschitz. The first result of this section is given below.

Theorem 1.10 *A subset $\mathcal{F}$ of $\mathcal{C}$ is precompact if and only if*

(1) *$D(\mathcal{F})$ is a precompact subset of $\mathcal{M}$.*
(2) *For every $\varepsilon > 0$, there is $\delta > 0$ such that if $f \in \mathcal{F}$, $x, x' \in D(f)$, and $d(x, x') < \delta$, then $d(f(x), f(x')) < \varepsilon$.*

For simplicity, a subset $\mathcal{F}$ of $\mathcal{C}$ satisfying (1) and (2) above will be referred to as an *equicontinuous family*.

This theorem is a counterpart of the Arzelà-Ascoli theorem [61]. Nonetheless, the convergence in [61] is different from the one obtained from d_{GH^0} (compare with [61, p. 401]).

This theorem is false if we consider families satisfying only (2) in the definition of equicontinuous family (take, for instance, the set of isometries). It is also false if we replace $\mathcal{C}$ by the set of homeomorphisms of compact metric spaces up to isometric conjugacy.

To state our second result, we say that $g \in \mathcal{C}$ is a *periodic orbit* if $D(g)$ consists of a single periodic orbit. We also say that g is *chain transitive* if for every $y, y' \in D(g)$ and every $\delta > 0$, there is a sequence $y_0 = y, y_1, \ldots, y_r = y'$ such that $d(f^k(y_k), y_{k+1}) \leq \delta$ for $0 \leq k \leq r - 1$.

In our second result, we characterize the chain transitive maps as those which can be C^0-Gromov-Hausdorff approximated by periodic orbits.

Theorem 1.11 $f \in \mathcal{C}$ is chain transitive if and only if there is a sequence of periodic orbits $\{g_n\}_{n \in \mathbb{N}}$ in $\mathcal{C}$ such that $g_n \xrightarrow{GH^0} f$.

This result was originally proved by Jung [40], motivated by the MSc dissertation by Cubas [24]. The proof we will give here is new.

In view of this result, it is natural to ask whether every continuous map that can be approximated in the C^0-Gromov-Hausdorff sense (GH^0-approximated for short) by periodic orbits is transitive, that is, has a dense orbit. However, the following counterexample shows that this is not the case.

Example 1.14 The identity map of S^1 is not transitive. However, it can be GH^0-approximated by periodic orbits $g_n : Y_n \to Y_n$. Indeed, it suffices to choose g_n to be a rational rotation on a finite set of S^1 such that its rotation number goes to 0 as $n \to \infty$.

Let us give a condition under which C^0-Gromov-Hausdorff approximately by periodic orbits implies transitivity. We say that a map $f : X \to X$ is *Lipschitz* if there is a constant $K > 1$ (called Lipschitz constant) such that $d(f(a), f(b)) \leq Kd(a, b)$ for all $a, b \in X$.

To simplify the notation, we shall denote by $X_n \xrightarrow{GH} X$ and $f_n \xrightarrow{GH^0} f$ (as $n \to \infty$) the convergences in $\mathcal{M}$ and $\mathcal{C}$, respectively. The convergence in other spaces will be denoted by $a_n \to a$.

Theorem 1.12 Let $f \in \mathcal{C}$ be Lipschitz with Lipschitz constant K and let $\{g_n : Y_n \to Y_n\}_{n \in \mathbb{N}}$ be a sequence of periodic orbits with periods $\pi(g_n)$. If

$$g_n \xrightarrow{GH^0} f \quad and \quad \pi(g_n) K^{\pi(g_n)} d_{GH^0}(f, g_n) \to 0 \quad as \quad n \to \infty,$$

then f is transitive.

To prove Theorem 1.10, we need two lemmas.

Lemma 1.11 *Every precompact subset of C satisfies (2) in the definition of an equicontinuous family.*

Proof Suppose by contradiction that there is $\mathcal{F} \subseteq C$ which is precompact, but does not satisfy (2) in the definition of an equicontinuous family. Then there are $\varepsilon > 0$ and sequences $\{f_n\}_{n\in\mathbb{N}}$ in $\mathcal{F}$, $\{x_n\}_{n\in\mathbb{N}}$ and $\{x'_n\}_{n\in\mathbb{N}}$ in $D(f_n)$ such that $d(x_n, x'_n) < \frac{1}{n}$, but $d(f(x_n), f(x'_n)) \geq \varepsilon$ for any $n \in \mathbb{N}$. Since $\mathcal{F}$ is precompact, we can assume that $f_n \xrightarrow{\mathrm{GH}^0} f$ for some $f \in C$. So, there are a sequence $\{\delta_n\}_{n\in\mathbb{N}}$ with $\delta_n \to 0^+$ and δ_n-isometries $j_n : D(f_n) \to D(f)$ such that $d_{C^0}(j_n \circ f_n, f \circ j_n) < \delta_n$ for every $n \in \mathbb{N}$. On the other hand, we have

$$d(f_n(x_n), f_n(x'_n)) < \delta_n + d(j_n(f_n(x_n)), j_n(f_n(x'_n)))$$
$$\leq \delta_n + d(j_n(f_n(x_n)), f(j_n(x_n))) + d(f(j_n(x_n)), f(j_n(x'_n)))$$
$$+ d(f(j_n(x'_n)), j_n(f_n(x'_n)))$$
$$\leq 3\delta_n + d(f(j_n(x_n)), f(j_n(x'_n))).$$

Also,

$$d(j_n(x_n), j_n(x'_n)) < \delta_n + d(x_n, x'_n)$$
$$\leq \delta_n + \frac{1}{n}.$$

So, $d(j_n(x_n), j_n(x'_n)) \to 0$ as $n \to \infty$, and consequently

$$d(f(j_n(x_n)), f(j_n(x'_n))) \to 0,$$

because f is continuous and $D(f)$ is compact. Therefore,

$$\varepsilon \leq d(f_n(x_n), f_n(x'_n)) \leq 3\delta_n + d(f(j_n(x_n)), f(j_n(x'_n))) \to 0,$$

which is absurd. This completes the proof.   □

Proof *(Of Theorem* 1.10*)* By Lemma 1.11, precompactness of $\mathcal{F}$ implies that $\mathcal{F}$ satisfies (2) in the definition of an equicontinuous family. But precompactness of $\mathcal{F}$ also implies (1) in the definition, namely, $D(\mathcal{F})$ is precompact in $\mathcal{M}$. Indeed, take a sequence $\{X_n\}_{n\in\mathbb{N}}$ in $D(\mathcal{F})$. Then $X_n = D(f_n)$ for some sequence $\{f_n\}_{n\in\mathbb{N}}$ in $\mathcal{F}$. Since $\mathcal{F}$ is precompact, there is a convergent subsequence, say $f_{n_l} \xrightarrow{\mathrm{GH}^0} f \in C$. Then,

$$d_{\mathrm{GH}}(X_{n_l}, D(f)) = d_{\mathrm{GH}}(D(f_{n_l}), D(f)) \leq d_{\mathrm{GH}^0}(f_{n_l}, f) \to 0.$$

So, $X_{n_l} \xrightarrow{\mathrm{GH}^0} D(f)$; thus, $D(\mathcal{F})$ is precompact. Therefore, $\mathcal{F}$ is equicontinuous.

Conversely, suppose that $\mathcal{F} \subseteq \mathcal{C}$ is equicontinuous. Take a sequence $\{f_n\}_{n \in \mathbb{N}} \subset \mathcal{F}$. We shall prove that f_n has a convergent subsequence. For this, we note that $D(f_n) \in D(\mathcal{F})$; so we can assume that $D(f_n) \xrightarrow{\mathrm{GH}} X$ for some $X \in \mathcal{M}$. From this and Lemma 1.6 we get sequences of δ_n-isometries $\{i_n : X \to D(f_n)\}_{n \in \mathbb{N}}$ and $\{j_n : D(f_n) \to X\}_{n \in \mathbb{N}}$ with $\delta_n \to 0$ as $n \to \infty$, such that

$$d(j_n \circ i_n, I_X) \to 0 \text{ and } d(i_n \circ j_n, I_{D(f_n)}) \to 0 \text{ as } n \to \infty.$$

Consider the composition

$$j_n \circ f_n \circ i_n : X \to X.$$

Take a countable dense subset $A \subseteq X$. By a diagonal argument, we can assume that the sequence $j_n \circ f_n \circ i_n(a)$ converges to some $f(a) \in X$ for all $a \in A$. This defines a map $f : A \to X$ by

$$f(a) = \lim_{n \to \infty} j_n \circ f_n \circ i_n(a).$$

We claim that f is uniformly continuous. Indeed, take $\varepsilon > 0$ and let $\delta > 0$ be given by the equicontinuity of $\mathcal{F}$ for $\frac{\varepsilon}{2}$. If $d(a, a') < \frac{\delta}{2}$ for some $a, a' \in A$, then

$$d(i_n(a), i_n(a')) < \delta_n + d(a, a') < \delta_n + \frac{\delta}{2} < \delta$$

for n sufficiently large. So,

$$d(j_n \circ f_n \circ i_n(a), j_n \circ f_n \circ i_n(a')) < \delta_n + d(f_n(i_n(a)), f_n(i_n(a')))$$

$$< \delta_n + \frac{\varepsilon}{2} < \varepsilon$$

for n sufficiently large. Taking $n \to \infty$ above, we get $d(f(a), f(a')) < \varepsilon$. Therefore, f is uniformly continuous.

It follows that f can be extended continuously to the closure $\overline{A} = X$. This yields a uniformly continuous map, still denoted by $f : X \to X$.

It remains to prove $f_n \xrightarrow{\mathrm{GH}^0} f$. This is done by using the following short lemmas.

Lemma 1.12 $j_n \circ f_n \circ i_n$ *converges pointwise to f as $n \to \infty$.*

Proof Fix $x \in X$ and $\varepsilon > 0$. Let $\bar{\delta}$ be given as in (2) of Theorem 1.10 for $\frac{\varepsilon}{4}$. Notice that A is dense in X, f is continuous, $j_n \circ f_n \circ i_n(a) \to f(a)$ for every $a \in A$, and $\delta_n \to 0$. Then there are $n_1 \in \mathbb{N}^+$ and $a \in A$ such that

$$\delta_n < \frac{\varepsilon}{12}, \quad d(f(a), f(x)) < \frac{\varepsilon}{3}, \quad d(j_n \circ f_n \circ i_n(a), f(a)) < \frac{\varepsilon}{3}$$

and

$$d(x, a) < \min\{\bar{\delta} - \delta_n, \frac{\varepsilon}{3}\}$$

for all $n \geq n_1$. Since i_n is a δ_n-isometry, $d(i_n(x), i_n(a)) < \delta_n + d(x, a) < \bar{\delta}$. So,

$$d(j_n \circ f_n \circ i_n(x), j_n \circ f_n \circ i_n(a)) < \delta_n + d(f_n(i_n(x)), f_n(i_n(a))) < \delta_n + \frac{\varepsilon}{4} < \frac{\varepsilon}{3}.$$

Thus,

$$d(j_n \circ f_n \circ i_n(x), f(x)) < d(j_n \circ f_n \circ i_n(x), j_n \circ f_n \circ i_n(a))$$
$$+ d(j_n \circ f_n \circ i_n(a), f(a)) + d(f(a), f(x))$$
$$< \frac{\varepsilon}{3} + \frac{\varepsilon}{3} + \frac{\varepsilon}{3} = \varepsilon$$

for all $n \geq n_1$, proving the result. $\square$

Lemma 1.13 $j_n \circ f_n \circ i_n$ *converges uniformly to* f *as* $n \to \infty$.

Proof Take $\varepsilon > 0$. Since both i_n and j_n are δ_n-isometries, $\delta_n \to 0$, and f is uniformly continuous, it follows from Item (2) in Theorem 1.10 that there are $\delta > 0$ and $n_1 \in \mathbb{N}$ such that

$$d(j_n \circ f_n \circ i_n(p), j_n \circ f_n \circ i_n(q)) < \frac{\varepsilon}{3} \quad \text{and} \quad d(f(p), f(q)) < \frac{\varepsilon}{3}$$

whenever $d(p, q) < \delta$ and $n \geq n_1$. Fix a finite cover

$$X = \bigcup_{l=1}^{r} B(x_l, \delta).$$

By Lemma 1.12, we can enlarge n_1 if necessary to get

$$d(f(x_l), j_n \circ f_n \circ i_n(x_l)) < \frac{\varepsilon}{3}$$

for $n \geq n_1$ and $1 \leq l \leq r$. Now fix $n \geq n_1$. If $x \in X$, then since we have the above δ-cover there is an l, $1 \leq l \leq r$, such that $d(x, x_l) < \delta$. Consequently,

$$d(f(x), j_n \circ f_n \circ i_n(x)) \leq d(f(x), f(x_l)) + d(f(x_l), j_n \circ f_n \circ i_n(x_l))$$

$$+ d(j_n \circ f_n \circ i_n(x_l), j_n \circ f_n \circ i_n(x))$$

$$< \frac{\varepsilon}{3} + \frac{\varepsilon}{3} + \frac{\varepsilon}{3} = \varepsilon.$$

We conclude that $d(j_n \circ f_n \circ i_n, f) < \varepsilon$ if $n \geq 1$, proving the lemma. $\qquad\square$

Now we use these lemmas to complete the proof of the theorem. Since

$$d(j_n \circ f_n \circ i_n \circ j_n, f \circ j_n) \leq d(j_n \circ f_n \circ i_n, f)$$

and $d(j_n \circ f_n \circ i_n, f) \to 0$ as $n \to \infty$ by Lemma 1.13, we get

$$d(j_n \circ f_n \circ i_n \circ j_n, f \circ j_n) \to 0 \text{ as } n \to \infty. \tag{1.9}$$

On the other hand, we leave that

$$d(f_n, f_n \circ i_n \circ j_n) \to 0. \tag{1.10}$$

Indeed, fix $\varepsilon > 0$ and let δ be given by (2) in Theorem 1.10 for this ε. Since $d(i_n \circ j_n, I_{D(f_n)}) \to 0$ as $n \to \infty$, we can choose $n_1 \in \mathbb{N}$ such that

$$d(i_n \circ j_n(y), y) < \delta$$

for $n \geq n_1$ and $y \in D(f_n)$. Therefore,

$$d(f_n(y), f_n(i_n \circ j_n(y))) < \varepsilon,$$

proving the assertion.
On the other hand,

$$d(j_n \circ f_n, f \circ j_n) \leq d(j_n \circ f_n, j_n \circ f_n \circ i_n \circ j_n) + d(j_n \circ f_n \circ i_n \circ j_n, f \circ j_n)$$

$$< \delta_n + d(f_n, f_n \circ i_n \circ j_n) + d(j_n \circ f_n \circ i_n \circ j_n, f \circ j_n) \to 0$$

as $n \to \infty$. So combining (1.9) and (1.10), we obtain

$$d(j_n \circ f_n, f \circ j_n) \to 0$$

as $n \to \infty$. Since j_n is a δ_n-isometry and $\delta_n \to 0^+$, Lemma 3.2 implies that

$$f_n \xrightarrow{\text{GH}^0} f,$$

completing the proof. $\square$

Proof (Of Theorem 1.11) First, we prove the necessity. By Lemma 3.2, it suffices to show that for every $\Delta' > 0$, there exist a periodic orbit g and a Δ'-isometry $j : D(g) \to D(f)$ such that $d(f \circ j, j \circ g) \leq \Delta'$. If $\Delta' > 0$, we can choose a finite set $F = \{q_1, \ldots, q_p\} \subset X$ such that $d(q_1, f(q_p)) < \Delta'$ and $d_H(F, X) < \Delta'$. Since f is chain transitive, there is a finite sequence $q_1 = x_0, x_1, \ldots, x_n = q_p$ of X such that $F \subset \{x_0, \ldots, x_n\}$ and $d(f(x_k), x_{k+1}) < \Delta'$ for $0 \leq k \leq n - 1$.

Let $Y = \{x_0, \ldots, x_{n-1}\}$ with the metric induced from X. Define $g : Y \to Y$ by $g(x_k) = x_{k+1}$ for $0 \leq k \leq n - 1$, $g(x_n) = x_0$ and $j : Y \to X$ as the inclusion.

Since j is the inclusion, $|d(j(y), j(y')) - d(y, y')| = 0 < \Delta'$ for every $y, y' \in Y$. Hence,

$$\sup_{y, y' \in Y} |d(j(y), j(y')) - d(y, y')| < \Delta'.$$

Also, if $x \in X$, then there is $y \in F$ such that $d(x, y) < \delta'$. As $F \subset Y$, we get $d_H(j(Y), X) < \Delta'$, proving that j is a Δ'-isometry.

Moreover, since

$$d(f \circ j(x_k), j \circ g(x_k)) = d(f(x_k), x_{k+1}) < \Delta'$$

for $0 \leq k \leq n - 1$ and

$$d(f \circ j(x_n), j \circ g(x_n)) = d(f(x_n), x_0) = d(f(q_p), q_1) < \Delta',$$

we get $d(f \circ j, j \circ g) < \Delta'$, as needed.

Now we prove the sufficiency. Suppose that there is a sequence of periodic orbits $\{g_n\}_{n \in \mathbb{N}}$ with $g_n \xrightarrow{\text{GH}^0} f$. By definition, we can take a sequence $\{\delta_n\}_{n \in \mathbb{N}}$ with $\delta_n \to 0^+$ and δ_n-isometries $j_n : Y_n \to X$ such that $d^X(j_n \circ g_n, f \circ j_n) < \delta_n$ for all $n \in \mathbb{N}$. Fix $\varepsilon > 0$ and $x, x' \in X$. Since j_n is a δ_n-isometry with $\delta_n \to 0$ and each g_n is a periodic orbit, there are $y_n \in Y_n$ and $0 \leq l_n \leq \pi(g_n)$ such that $d(x, j_n(y_n)) < \varepsilon$ and $d(x', j_n(g_n^{l_n}(y_n))) < \varepsilon$ for all n large enough.

Now consider the sequence $x_1, \ldots, x_{l_n} \in X$ defined by $x_m = j_n(g_n^m(y_n))$ for $1 \leq m \leq l_n$. Since

$$d(f(x_m), x_{m+1}) = d(f(j_n(g_n^m(y_n))), j_n(g_n^{m+1}(y_n)))$$

$$= d(f \circ j_n(g_n^m(y_n)), j_n \circ g_n(g_n^m(y)))$$

$$< \delta,$$

we get that the above sequence is a δ_n-chain. As $\delta_n \to 0$, f is chain transitive and the result follows. □

To prove Theorem 1.12, we will use the following short lemma.

Lemma 1.14 *Let* $f \in C$ *and* $\{g_n\}_{n \in \mathbb{N}}$ *be a sequence in* C. *If* $a_n > 0$ *is a sequence such that* $\lim_{n \to \infty} a_n d_{GH^0}(g_n, f) \to 0$, *then there are sequences* $\{k_n\}_{n \in \mathbb{N}} \subset \mathbb{N}$, $\{\delta_{k_n}\}_{n \in \mathbb{N}}$ *with* $\delta_{k_n} \to 0^+$, *and* δ_{k_n}-*isometries* $j_{k_n} : Y_{k_n} \to X$ *such that*

$$\lim_{n \to \infty} a_{k_n} \delta_{k_n} = 0.$$

Proof Given $n \in \mathbb{N}$, there is $k_n \in \mathbb{N}$ such that $a_{k_n} d_{GH^0}(g_{k_n}, f) < \frac{1}{n}$. Then $d_{GH^0}(g_{k_n}, f) < \frac{1}{a_{k_n} n}$; so there is $d_{GH^0}(g_{k_n}, f) < \delta_{k_n} < \frac{1}{a_{k_n} n}$. It follows that $\lim_{n \to 0} a_{k_n} \delta_{k_n} = 0$ and we get a δ_{k_n}-isometry $j_{k_n} : Y_{k_n} \to X$ from the definition of d_{GH^0}. □

Proof (Of Theorem 1.12) To prove that f is transitive, it suffices to show that for every pair of open sets $U, V \subset X$, there is $l \in \mathbb{N}$ such that $f^l(U) \cap V \neq \emptyset$. Given such U and V, we will find l as follows.

Put $a_n = \pi(g_n) K^{\pi(g_n)}$ as in Lemma 1.14 to get sequences $\{k_n\}_{n \in \mathbb{N}} \subset \mathbb{N}$, $\{\delta_{k_n}\}_{n \in \mathbb{N}} \to 0$, and δ_{k_n}-isometries $j_{k_n} : Y_{k_n} \to X$. For simplicity, we assume $k_n = n$ for all $n \in \mathbb{N}$.

We will prove

$$d(f^l(j_n(y_n)), j_n(g_n^l(y_n))) \leq l K^l \delta_n \qquad (1.11)$$

for all $y_n \in Y_n$ and $1 \leq l \leq \pi(g_n)$. Indeed, since $d(f(j_n(y_n)), j_n(g_n(y_n))) = d(f \circ j_n(y_n), j_n \circ g_n(y_n)) \leq \delta_n$, we have

$$d(f(j_n(y_n)), j_n(g_n(y_n))) \leq \delta_n \leq K\delta_n,$$

proving (1.11) for $l = 1$.

Now suppose that $1 \leq l < \pi(g_n)$ satisfies (1.11). Since

$$d(f^{l+1}(j_n(y_n)), j_n(g_n^{l+1}(y_n))) \leq d(f^l(f \circ j_n(y_n)), f^l(j_n \circ g_n(y_n)))$$

$$+ d(f^l(j_n(g_n(y_n))), j_n(g_n^l(g_n(y_n))))$$

$$\leq K^l d(f \circ j_n(y_n), j_n \circ g_n(y_n)) + l K^l \delta_n$$

$$\leq K^l \delta_n + l K^l \delta_n$$

$$= (l+1) K^l \delta_n,$$

(1.11) holds for $l+1$. Therefore, (1.11) is proved by induction. Now we use (1.11) to complete the proof of the theorem.

Fix an open set $V' \subset V$ with closure $\overline{V'} \subset V$. Since $j_n : Y_n \to X$ is a δ_n-isometry, $d_H(j_n(Y_n), X) \leq \delta_n$. So, $j_n(Y_n) \cap U \neq \emptyset$ for all large enough n. Hence, we can choose a sequence $\{y_n\}_{n \in \mathbb{N}} \subset Y_n$ such that $j_n(y_n) \in U$ for all large enough n. Likewise, there is a sequence $\{y'_n\}_{n \in \mathbb{N}} \subset Y_n$ with $y'_n \in V'$ for every n large. Since g_n is periodic, $y'_n = g_n^{l_n}(y_n)$ for some sequence $\{l_n\}_{n \in \mathbb{N}} \subset \mathbb{N}$. Thus, $j_n(y_n) \in U$ and $j_n(g_n^{l_n}(y_n)) \in V'$ for all enough n. Since $K > 1$, by applying (1.11), we obtain

$$d(f^{l_n}(j_n(y_n)), j_n(g_n^{l_n}(y_n))) \leq l_n K^{l_n} \delta_n \leq \pi(g_n) K^{\pi(g_n)} \delta_n$$

for $n \in \mathbb{N}$. Since $a_n = \pi(g_n) K^{\pi(g_n)}$, we have $\pi(g_n) K^{\pi(g_n)} \delta_n \to 0$ as $n \to \infty$ by Lemma 1.14. Hence,

$$d(f^{l_n}(j_n(y_n)), j_n(g_n^{l_n}(y_n))) \to 0$$

as $n \to \infty$. Since $j_n(g_n^{l_n}(y_n)) \in V'$ and $\overline{V'} \subset V$, we get $f^{l_n}(j_n(y_n)) \in V$ for n large enough. Taking $l = l_n$ and $x = j_n(y_n)$, we get $f^l(x) \in f^l(U) \cap V$, and so $f^l(U) \cap V \neq \emptyset$. This completes the proof. □

Exercises

Exercise 1.15 Prove that a surjective map $i : X \to Y$ of compact metric spaces is an isometry if and only if $d(i(x), i(x')) \leq d(x, x')$ for every $x, x' \in X$ (see [26] for the case $X = Y$).

Exercise 1.16 Prove that a complete metric space is compact if and only if it has a finite $\frac{1}{n}$-net for every $n \in \mathbb{N}$. Prove also that a Cauchy sequence in a metric space is convergent if and only if it has a convergent subsequence.

Exercise 1.17 Prove that the Gromov-Hausdorff space is noncompact and path connected.

Exercise 1.18 (Original Gromov-Hausdorff distance) Let X and Y be compact metric spaces. Define $\hat{d}_{GH}(X, Y)$ as the infimum of all $\Delta > 0$ such that there exist a metric space Z and isometric immersions $i : X \to Z$ and $j : Y \to Z$ such that

$d(i(X), j(Y)) \leq \Delta$. Prove that $\hat{d}_{GH}$ is equivalent to d_{GH}, namely, that there is $K > 1$ such that $K^{-1}d_{GH}(X, Y) \leq \hat{d}_{GH}(X, Y) \leq Kd_{GH}(X, Y)$ for all $X, Y \in M$.

Exercise 1.19 Prove that the finite metric spaces form a dense subset of the Gromov-Hausdorff space. Conclude that the Gromov-Hausdorff space is separable.

Exercise 1.20 Prove the necessity in Gromov precompactness theorem, namely, slow that if $\mathcal{X} \subset \mathcal{M}$ is precompact, then

1. $\sup_{X \in \mathcal{X}} \text{diam}(X) < \infty$;
2. for every $\varepsilon > 0$, there is $N(\varepsilon) \in \mathbb{N}$ such that every $X \in \mathcal{X}$ has an ε-net of at most than $N(\varepsilon)$ elements.

Exercise 1.21 Find continuous maps $f, g : X \to X$ of a compact metric space such that $d_{GH^0}(f, g) < d(f, g)$ (compare with Item (1) of Theorem 1.9).

Exercise 1.22 Find continuous maps $f : X \to X$, $g : Y \to Y$ and $r : Z \to Z$ of compact metric spaces such that $d_{GH^0}(f, g) > d_{GH^0}(f, r) + d_{GH^0}(r, g)$.

Exercise 1.23 Prove that if $f : X \to X$ and $g : Y \to Y$ are continuous maps of compact metric spaces, then

$$d_{GH^0}(f, g) \leq \max\{d_{GH}(X, Y), \text{diam}(X), \text{diam}(Y)\}.$$

Exercise 1.24 Let $f : X \to X$ be a continuous map of a compact metric space. Suppose that for every $\Delta > 0$, there is $\Delta' > 0$ such that if $g : Y \to Y$ is a continuous map of a compact metric space satisfying $d(g \circ i, i \circ f) \leq \Delta'$ for some Δ'-isometry $i : X \to Y$, then $d_{GH^0}(f, g) \leq \Delta$.

Exercise 1.25 Prove the following properties of the Gromov-Hausdorff distance for group actions $d_{GH^0,A}(T, S)$, where G is a group with generating set A:

1. $D_{GH^0}(X, Y) \leq d_{GH^0,A}(T, S)$.
2. If $X = Y$, then

$$d_{GH^0,A}(T, S) \leq \sup_{a \in A, x \in X} d(T_a(x), S_a(x)).$$

3. $d_{GH^0,A}(T, S) < \infty$.
4. If $R \in \text{Act}(G, Z)$ for a compact metric space Z, then

$$d_{GH^0,A}(T, S) \leq 2(d_{GH^0,A}(T, R) + d_{GH^0,A}(R, S)).$$

5. If A is finite, then $d_{GH^0,A}(T, S) = 0$ if and only if T and S are isometric.

Exercise 1.26 Is the GH-variational principle (Theorem 1.6) valid for every pair of compact metric spaces X and Y?

A map $i : X \to Y$ of compact metric spaces is δ-*continuous* if there is $\beta > 0$ such that $d(x, x') \leq \beta$ implies $d(i(x), i(x')) \leq \delta$; δ-*surjective* if $d_H(i(X), Y) \leq \delta$; and δ-*injective* if there is $\beta > 0$ such that $d(i(x), i(x')) \leq \beta$ implies $d(x, x') \leq \delta$. We say that i is a δ-*homeomorphism* if it is δ-continuous, δ-surjective and δ-injective.

Exercise 1.27 Define $D : M \times M \to [0, \infty[$ by setting $D(X, Y)$ as the infimum of the numbers $\delta > 0$ such that there are δ-homeomorphisms $i : X \to Y$ and $j : Y \to X$. Prove that this D is normalized and so if does not satisfy *(P3)* in Sect. 1.6. Does $D(X, Y) = 0$ imply that X and Y are homeomorphic?

Exercise 1.28 A continuous map $f : X \to X$ of a compact metric space is *minimal* if the sets $\{f^n(x) : n \in \mathbb{N}\}$ are dense in X for every $x \in X$. Find an example of a homeomorphism of a compact metric space that is not transitive but can be GH^0-approximated by minimal homeomorphisms.

Exercise 1.29 Let $\{g_n\}_{n \in \mathbb{N}} \subset C$ be a sequence such that $g_n \xrightarrow{GH^0} f$ for some $f \in C$. Suppose that each g_n has a compact invariant set Λ_n. Is it true that there is a compact invariant set Λ of f such that $\Lambda_n \xrightarrow{GH} \Lambda$?

Exercise 1.30 A continuous map $f : X \to X$ of a compact metric space is *equicontinuous* if for every $\varepsilon > 0$, there is $\delta > 0$ such that $x, x' \in X$ and $d(x, x') \leq \delta$ implies $d(f^n(x), f^n(x')) \leq \varepsilon$ for every $n \in \mathbb{N}$. Is the C^0-Gromov-Hausdorff limit of a sequence of equicontinuous maps equicontinuous?

Exercise 1.31 Is C complete? Try to use Theorem 1.10 to prove it.

Exercise 1.32 Let $\mathcal{H}$ be the space of isometric classes of homeomorphisms of compact metric spaces. Is $\mathcal{H}$ complete?

Exercise 1.33 Let $D : C \to M$ be the map assigning to each $f \in C$ its domain $D(f)$ (up to isometries). Prove that D is well-defined and

$$D_{GH^0}(D(f), D(g)) \leq d_{GH^0}(f, g) \qquad \forall f, g \in C.$$

Exercise 1.34 Prove that the map D in Exercise 1.33 has a *global cross section*, that is, there is $C' \subset C$ such that D maps C' homeomorphically onto M. Is $D : C \to M$ a covering map or a fibre bundle?

Exercise 1.35 Use Lemma 1.13, (1.9), (1.10) and Exercise 1.24 to prove Theorem 1.10.

Exercise 1.36 Extend and prove the results in Sect. 1.8 to finitely generated groups actions, flows or semiflows.

Stability

2

2.1 Introduction

In this chapter, we use the C^0-Gromov-Hausdorff distance between maps of metric spaces to introduce the notion of *topological* GH-*stability*. We will prove that there are topologically stable homeomorphisms that are not topologically GH-stable. Also, we show that every topologically GH-stable circle homeomorphism is topologically stable. We then prove that every expansive homeomorphism with the shadowing property of a compact metric space is topologically GH-stable. This is related to Walters' stability theorem [76]. We extend these results to group actions and to dissipative semiflows on complete metric spaces.

2.2 Definitions and Statement of Main Results

We first recall the classical notion of topological stability introduced by Walters [75]. Recall that Id denotes the identity map of a metric space X (sometimes we write Id_X instead to emphasize the dependence on X).

Definition 2.1 A homeomorphism $f : X \to X$ of a compact metric space X is *topologically stable* if for every $\varepsilon > 0$, there is $\delta > 0$ such that for every homeomorphism $g : X \to X$ with $d(f, g) < \delta$, there is a continuous map $h : X \to X$ with $d(h, \mathrm{Id}_X) < \varepsilon$ such that $f \circ h = h \circ g$.

This definition together with the C^0-Gromov-Hausdorff distance motivates the following notion of stability.

© The Author(s), under exclusive license to Springer Nature Switzerland AG 2022
J. Lee, C. Morales, *Gromov-Hausdorff Stability of Dynamical Systems and Applications to PDEs*, Frontiers in Mathematics, https://doi.org/10.1007/978-3-031-12031-2_2

Definition 2.2 A homeomorphism $f : X \to X$ of a compact metric space X is *topologically GH-stable* if for every $\varepsilon > 0$, there is $\delta > 0$ such that for every homeomorphism $g : Y \to Y$ of a compact metric space Y satisfying $d_{GH^0}(f, g) < \delta$, there is a continuous ε-isometry $h : Y \to X$ such that $f \circ h = h \circ g$.

Both definitions can be extended to continuous maps by just replacing homeomorphism by continuous map in their statement.

In our first result, we will prove that there are topologically stable homeomorphisms that are not topologically GH-stable.

Theorem 2.1 *There exist a compact metric space X and a homeomorphism $f : X \to X$ that is topologically stable, but not topologically GH-stable.*

Although we do not know if every topologically GH-stable homeomorphism is topologically stable, we will prove that this is the case for circle homeomorphisms.

Theorem 2.2 *Every topologically GH-stable circle homeomorphism is topologically stable.*

Next, we recall that a homeomorphism $f : X \to X$ of a metric space X is *expansive* if there is $\delta > 0$ (called the *expansivity constant*) such that if $x, y \in X$ satisfy $d(f^n(x), f^n(y)) \leq \delta$ for all $n \in \mathbb{Z}$, then $x = y$. Moreover, f has the *shadowing property* if for every $\varepsilon > 0$, there is $\delta > 0$ such that for every bi-infinite sequence $\{x_n\}_{n \in \mathbb{Z}}$ satisfying $d(x_{n+1}, f(x_n)) < \delta$ for every $n \in \mathbb{Z}$, there is $x \in X$ such that $d(f^n(x), x_n) < \varepsilon$ for every $n \in \mathbb{Z}$.

Walters' stability theorem asserts that every expansive homeomorphism with the shadowing property of a compact metric space is topologically stable [76]. Here, we will prove that all such homeomorphisms are also topologically GH-stable.

Theorem 2.3 *Every expansive homeomorphism with the shadowing property of a compact metric space is topologically GH-stable.*

2.3   Proof of Theorem 2.1

The proof will use the following lemma. Recall that a homeomorphism $g : X \to X$ of a metric space X is *minimal* if $O_g(x)$ is dense in X for all $x \in X$, where $O_g(x) = \{g^n(x) : n \in \mathbb{N}\}$ is the g-orbit of x.

Lemma 2.1 *Let $f : X \to X$ be a topologically GH-stable homeomorphism of a compact metric space X. If $\inf_{z \in X} d_H(X, O_f(z)) > 0$, then there is $\delta > 0$ such that* *no homeomorphism* $g : Y \to Y$ *of a compact metric space* Y *with* $d_{GH^0}(f, g) < \delta$ *is minimal.*

Proof By the hypothesis, there is $\varepsilon > 0$ such that

$$d_H(X, O_f(z)) > \varepsilon$$

for all $z \in X$. For this ε, we let $\delta > 0$ be given by the topological GH-stability of f. If this δ does not work, then we could select a minimal homeomorphism $g : Y \to Y$ of a compact metric space Y with $d_{GH^0}(f, g) < \delta$. Then by topological GH-stability, there is a continuous ε-isometry $h : Y \to X$ such that $f \circ h = h \circ g$.

Fix $y \in Y$. Since g is minimal, $O_g(y)$ is dense in Y and so $h(O_g(y))$ is dense in $h(Y)$ by the continuity of h. On the other hand, $f \circ h = h \circ g$. So, $h(O_g(y)) = O_f(z)$, where $z = h(y)$. It follows that $O_f(z)$ is dense in $h(Y)$. But h is an ε-isometry, so $d_H(X, h(Y)) < \varepsilon$. Since $O_f(z)$ is dense in $h(Y)$, we conclude that $d_H(X, O_f(z)) < \varepsilon$, contradicting the choice of ε. This completes the proof. $\qquad\qquad\qquad\qquad\qquad\qquad\square$

Proof (*Proof of Theorem* **2.1**) Consider the unforced undamped *Duffing oscillator* given by the flow Φ in $\mathbb{R}^2$ generated by the system

$$\begin{cases} \dot{x} = y, \\ \dot{y} = x - x^3. \end{cases}$$

It has three types of orbits: the outer circles Y, the ones forming the figure eight set X (which contains the equilibrium p), and two inner circles. We define $f : X \to X$ as the time-one map Φ_1 of Φ restricted to X.

Let us prove first that f is topologically stable.

Split $X = C_1 \cup C_2$ as the union of two circles C_1, C_2, intersecting at the equilibrium p. Clearly, any homeomorphism $g : X \to X$ fixes p, and if g is C^0-close to f, then it also leaves C_1 and C_2 invariant.

Take $\varepsilon > 0$. It follows that there is $\delta > 0$ such that if $d_{C^0}(f, g) < \delta$, then there are four fixed points $a_1, b_1 \in C_1$ and $a_2, b_2 \in C_2$ satisfying the following properties:

- $\max\{d(p, a_1), d(p, b_1), d(p, a_2), d(p, b_2)\} < \varepsilon$.
- $g([p, a_1] \cup [p, b_1] \cup [p, a_2] \cup [p, b_2]) = [p, a_1] \cup [p, b_1] \cup [p, a_2] \cup [p, b_2]$.
- If $x \in C_1 \setminus ([p, a_1] \cup [p, b_1])$, then $\lim_{n \to \infty} g^n(x) = b_1$ and $\lim_{n \to -\infty} g^n(x) = a_1$.
- If $x \in C_2 \setminus ([p, a_2] \cup [p, b_2])$, then $\lim_{n \to \infty} g^n(x) = a_2$ and $\lim_{n \to -\infty} g^n(x) = b_2$.

For such maps g, we pick $e_i \in C_i$ ($i = 1, 2$) and consider the intervals $[e_i, g(e_i)[$ in C_i for $i = 1, 2$. It follows that for every $x \in C_i \setminus ([p, a_i] \cup [p, b_i])$, there is a unique integer $n_i(x)$ such that $g^{-n_i(x)}(x) \in [e_i, g(e_i)[$ for $i = 1, 2$.

Making δ smaller if necessary, we obtain diffeomorphisms $h_i : [e_i, g(e_i)[\to [e_i, f(e_i)[$ that is C^0-close to the identity for $i = 1, 2$. Define $h : X \to X$ by

$$h(x) = \begin{cases} f^{n_1(x)}(h_1(g^{-n_1(x)}(x))), & \text{if } x \in C_1 \setminus ([p, a_1] \cup [p, b_1]), \\ f^{n_2(x)}(h_2(g^{-n_2(x)}(x))), & \text{if } x \in C_2 \setminus ([p, a_2] \cup [p, b_2]), \\ p, & \text{if } x \in [p, a_1] \cup [p, b_1] \cup [p, a_2] \cup [p, b_2]. \end{cases}$$

Straightforward computations show that h is continuous and $d_{C^0}(h, \mathrm{Id}_X) < \varepsilon$. Moreover, if $x \in C_1 \setminus ([p, a_1] \cup [p, b_1])$, then $f(h(x)) = f^{n_1(x)+1}(h_1(g^{-n_1(x)}(x)))$. But $n_1(g(x)) = n_1(x) + 1$; so, $f(h(x)) = f^{n_1(g(x))}(h_1(g^{-n_1(g(x))}(g(x)))) = h(g(x))$. Hence, $f(h(x)) = h(g(x))$ if $x \in C_1 \setminus ([p, a_1] \cup [p, b_1])$. Similarly, we can prove that $f(h(x)) = h(g(x))$ if $x \in C_2 \setminus ([p, a_2] \cup [p, b_2])$.

Next, if $x \in [p, a_1] \cup [p, b_1] \cup [p, a_2] \cup [p, b_2]$, then $f(h(x)) = f(p) = p$ and $h(g(x)) = p$, since $g([p, a_1] \cup [p, b_1] \cup [p, a_2] \cup [p, b_2]) = [p, a_1] \cup [p, b_1] \cup [p, a_2] \cup [p, b_2]$. All together, we have $f \circ h = h \circ g$. Therefore, f is topologically stable.

Now we prove that f is not topologically GH-stable. Take $g : Y \to Y$ as the time-one map $\Phi_1|_Y$ of Φ restricted to the outer circle Y. One can easily check that $d_{GH^0}(f, g) \to 0$ as Y converges to X. Also, since $D\Phi_1(p) \cdot \Phi(p) = \Phi(\Phi_1(p))$ for $p \in \mathbb{R}^2$, the map g is topologically conjugated to a circle rotation.

On the other hand, by taking Y close to X, we can see that the period of Y goes to infinity. Since this a period depends continuously on Y, we can choose Y arbitrarily close to X so that g has an irrational rotation number (and so it is minimal). Consequently, f can be C^0-GH-approximated by minimal homeomorphisms. Since f clearly satisfies the condition $\inf_{z \in X} d_H(X, O_f(z)) > 0$, f is not topologically GH-stable by Lemma 2.1. $\square$

2.4   Isometric Stability: Proof of Theorem 2.4

Item (1) of Theorem 1.9 implies that every topologically GH-stable homeomorphism of a compact metric space satisfies the auxiliary definition below.

Definition 2.3 We say that a homeomorphism $f : X \to X$ of a compact metric space X is *isometrically stable* if for every $\varepsilon > 0$, there is $\delta > 0$ such that for every homeomorphism $g : X \to X$ with $d(f, g) < \delta$, there is a *continuous ε-isometry* $h : X \to X$ satisfying $f \circ h = h \circ g$.

Since being ε-C^0-close to the identity implies being a 2ε-isometry, every topologically stable homeomorphism is isometrically stable. But it seems that the converse is false in general because an ε-isometry (or even an isometry) need not be ε-C^0-close to the identity. Nevertheless, we will prove such a converse in the case of the circle S^1.

Theorem 2.4 *A circle homeomorphism is isometrically stable if and only if it is topologically stable.*

Proof As noted above, every topologically stable homeomorphism is isometrically stable (not only on the circle, but also on every compact metric space). To prove the converse on S^1, we will follow the arguments in [78].

Recall that p is a periodic point of f if there is a minimal integer $n > 0$ (called a period) such that $f^n(p) = p$. Denote by $\mathrm{Per}(f)$ the set of periodic points of f. We say that $p \in \mathrm{Per}(f)$ is *topologically hyperbolic* if the map $f^{2n}(x) - x$ changes its sign at $x = p$, where n is the period of p.

As pointed by Yano [78], to prove that a homeomorphism $f : S^1 \to S^1$ is topologically stable, it suffices to prove that:

(a) $\mathrm{Per}(f)$ is non-empty and finite.
(b) Every element of $\mathrm{Per}(f)$ is topologically hyperbolic.

Hence, it suffices to prove that every isometrically stable homeomorphism $f : S^1 \to S^1$ satisfies (a) and (b).

Replacing f by f^2 if necessary, we can assume that f is orientation preserving.

Denote by $\mathrm{length}(I)$ the length of an interval $I \subset S^1$. Fix $0 < \varepsilon < \frac{1}{4}$ such that $\mathrm{length}(f(B)) < \frac{1}{8}$ whenever B is an interval of $\mathrm{length}(B) < 4\varepsilon$.

For this ε, we fix $0 < \delta < \frac{1}{16}$ from the isometric stability of f.

It follows from Peixoto's Theorem [55, p. 51] that there exists a diffeomorphism g with $d_{C^0}(f, g) < \delta$ such that $\mathrm{Per}(g)$ is non-empty and finite. Also, by the choice of δ, there is a continuous ε-isometry $h : S^1 \to S^1$ such that $f \circ h = h \circ g$.

Since $0 < \varepsilon < \frac{1}{4}$, we have that h is onto.

Indeed, denote by N, S, E, and W the north, south, east, and west poles on S^1. Denote by $[N, E]$ the interval in S^1 bounded by N and E, containing neither S nor W. Then $h([N, E])$ does not intersect $\{h(S), h(W)\}$. Otherwise, there would exist $P \in [N, E]$ with, say $h(P) = h(S)$, whence

$$\frac{1}{4} \leq d(P, S) = |d(h(P), h(S)) - d(P, S)| < \varepsilon < \frac{1}{4},$$

which is absurd. Similarly, the image of the corresponding intervals $[E, S]$, $[S, W]$, and $[W, N]$ under h does not intersect $\{h(N), h(W)\}$, $\{h(E), h(N)\}$, and $\{h(S), h(E)\}$, respectively. All together, we have that h is onto.

Next we take $x \in \mathrm{Per}(g)$ with period n. Since $f^n(h(x)) = h(g^n(x)) = h(x)$, we have $h(x) \in \mathrm{Per}(f)$, and so $\mathrm{Per}(f)$ is non-empty. Take any $x \in \mathrm{Per}(f)$ and consider $K = h^{-1}(O_f(x))$. Since h is onto, K is non-empty. If $y \in K$, then $h(y) = f^i(x)$ for some $0 \leq i \leq n - 1$ it is specified above. Thus, $h(g(y)) = f(h(x)) = f^{i+1}(x) \in O_f(x)$, proving $g(y) \in K$ whenever $y \in K$. Hence, K is compact, g-invariant, and non-empty.

It follows that K contains a periodic point of g. Since the collection $\{h^{-1}(O_f(x)) : x \in \text{Per}(f)\}$ is disjoint and $\text{Per}(g)$ is finite, we conclude that $\text{Per}(f)$ is finite. This proves (a).

The proof of (b) follows from the following

Claim: There is $\Delta > 0$ such that for every homeomorphism g with $d_{C^0}(f, g) < \Delta$ the cardinality of $\text{Per}(g)$ is not less than that of $\text{Per}(f)$.

Indeed, by (a), there is a positive integer n such that $\text{Per}(f) = \text{Fix}(f^n)$. Then, without loss of generality, we can assume that $\text{Per}(f) = \text{Fix}(f)$, where $\text{Fix}(f)$ is the set of fixed points of f. Now take $\Delta = \delta$ (where δ is as above), $x \in \text{Fix}(f)$, and a homeomorphism g with $d_{C^0}(f, g) < \delta$. By the choice of δ, unambiguously g is orientation-preserving. Also, there is a continuous ε-isometry $h : S^1 \to S^1$ with $f \circ g = g \circ h$. It follows that $h^{-1}(x)$ is a compact g-invariant set that is non-empty since h is onto. If $y, y' \in h^{-1}(x)$, then $h(y) = h(y') = x$. Then $d(y, y') = |d(h(y), h(y')) - d(y, y')| < \varepsilon$. Hence, $h^{-1}(x)$ is contained in an interval of length at most 2ε. It follows that, we can define $\sup h^{-1}(x)$ and $\inf h^{-1}(x)$ unambiguously. Putting $B = [\inf h^{-1}(x), \sup h^{-1}(x)]$, we have $\text{length}(B) \le 2\varepsilon$; so $\text{length}(f(B)) \le \frac{1}{8}$ by the choice of ε. Further, since $\text{length}(g(B)) \le \text{length}(f(B)) + 2d_{C^0}(f, g)$, we have

$$\text{length}(g(B)) \le \frac{1}{8} + 2\delta < \frac{1}{4}.$$

On the other hand, $h^{-1}(x)$ is g-invariant, so both endpoints of $g(B)$ are in B. Since g is a homeomorphism and the total length of S^1 is one, the estimate $\text{length}(g(B)) < \frac{1}{4}$ above implies that $g(B) \subset B$. Since B is an interval and g is continuous, this implies that g has a fixed point in B. Thus, we proved that each $x \in \text{Fix}(f)$ corresponds to a fixed point of g. So, whenever g satisfies $d_{C^0}(f, g) < \delta$, the cardinality of $\text{Per}(g)$ is at least that of $\text{Per}(f)$. The Claim is proved.

Now we use the Claim to prove (b). Suppose by contradiction that f has a periodic point that is not topologically hyperbolic. Then we can eliminate it by a small perturbation g, contradicting the Claim. Hence, (b) holds.   □

Another consequence of Theorem 2.4 is as follows. A diffeomorphism $f : S^1 \to S^1$ is called *Morse-Smale* if $\text{Per}(f)$ is non-empty and every element $x \in \text{Per}(f)$ satisfies $(f^n)'(x) \ne \pm 1$, where n is the period of x.

Corollary 2.1 *The following properties are equivalent for every homeomorphism $f : S^1 \to S^1$:*

(1) f is isometrically stable.
(2) f is topologically stable.
(3) f is topologically conjugate to a Morse-Smale diffeomorphism.

Proof (1) is equivalent to (2) by Theorem 2.4 and (2) is equivalent to (3) by Yano [78].

□

One more corollary is that, on in the circle, the isometric stability is invariant under topological equivalence. More precisely, we have the following result.

Corollary 2.2 *A circle homeomorphism topologically conjugate to an isometrically stable circle homeomorphism is isometrically stable.*

Proof Let f' be a circle homeomorphism that is topologically conjugate to an isometrically stable circle homeomorphism f. Since f is isometrically stable, by Corollary 2.1, there is a Morse-Smale diffeomorphism $g : S^1 \to S^1$ such that f and g are topologically conjugate. Since f and f' are homeomorphic, f' and g are also topologically conjugate. Then, f' is isometrically stable by Corollary 2.1.

□

We do not know if the isometric stability is invariant under topological equivalence on general metric spaces (the answer, however, seems to be negative). Meanwhile, let us recall that, in general, topological stability is invariant under topological conjugacy. An analogous result for topological GH-stability reads as follows.

Theorem 2.5 *Every homeomorphism of a compact metric space that is isometric to a topologically GH-stable homeomorphism is itself topologically GH-stable.*

Proof Let $f : X \to X$ and $f' : X' \to X'$ be homeomorphisms of compact metric spaces X and X'. Suppose that f and f' are isometric and that f is topologically GH-stable. Fix an isometry $v : X' \to X$ such that $f = v \circ f' \circ v^{-1}$. Take $\varepsilon > 0$ and let $\delta > 0$ be given by the topological GH-stability of f. Let $g' : Y' \to Y'$ be a homeomorphism of a compact metric space Y' such that $d_{\mathrm{GH}^0}(f', g') < \frac{\delta}{2}$. By Theorem 1.9,

$$d_{\mathrm{GH}^0}(f, g') \le 2(d_{\mathrm{GH}^0}(f, f') + d_{\mathrm{GH}^0}(f', g')) < 2 \cdot \frac{\delta}{2} = \delta.$$

Then, by the choice of δ, there is a continuous ε-isometry $h : Y' \to X$ such that $f \circ h = h \circ g'$. So, $v \circ f' \circ v^{-1} \circ h = h \circ g'$. Therefore, setting $h' = v^{-1} \circ h$, we get a continuous ε-isometry $h' : Y' \to X'$ satisfying $f' \circ h' = h' \circ g'$. Then f' is topologically GH-stable.

□

Proof (Of Theorem 2.2) Apply Theorem 2.4 and the fact that every topologically GH-stable homeomorphism is isometrically stable.

□

2.5   Proof of Theorem 2.5

The proof is inspired by Walters' stability theorem [76]. Let $f : X \to X$ be an expansive homeomorphism with the shadowing property of a compact metric space X. Fix $\varepsilon > 0$ and take $0 < \bar{\varepsilon} < \frac{1}{8}\min\{\varepsilon, e\}$, where e is the expansivity constant of f. For this $\bar{\varepsilon}$, we choose δ from the shadowing property. We can assume that $\delta < \bar{\varepsilon}$.

Now take a homeomorphism $g : Y \to Y$ of a compact metric space Y such that $d_{\mathrm{GH}^0}(f, g) < \delta$. Then there are δ-isometries $i : X \to Y$ and $j : Y \to X$ such that $d(g \circ i, i \circ f) < \delta$ and, more importantly,

$$d(j \circ g, f \circ j) < \delta.$$

Take $y \in Y$ and consider the sequence $\{x_n\}_{n \in \mathbb{Z}}$ defined by $x_n = j(g^n(y))$ for $n \in \mathbb{Z}$. Since

$$d(x_{n+1}, f(x_n)) = d(j(g(g^n(y))), f(j(g^n(y))))$$
$$= d(j \circ g(g^n(y)), f \circ j(g^n(y))) < \delta$$

for $n \in \mathbb{N}$, the choice of δ and the shadowing property provide a point $x \in X$ such that

$$d(f^n(x), x_n) \leq \bar{\varepsilon}.$$

In particular, $d(f^n(x), x_n) < \frac{e}{2}$ for all $n \in \mathbb{Z}$. Then such an x is unique (by expansivity). Denoting $x = h(y)$, we obtain a map $h : Y \to X$ satisfying

$$d(f^n(h(y)), j(g^n(y))) \leq \bar{\varepsilon}$$

for $y \in Y$ and $n \in \mathbb{Z}$. Taking here $n = 0$ above, we get $d(h(y), j(y)) < \varepsilon$ for every $y \in Y$, and so namely,

$$d(h, j) \leq \bar{\varepsilon}.$$

Then we have

$$d_H(h(Y), X) \leq d_H(h(Y), J(Y)) + d_H(j(Y), X) \leq \bar{\varepsilon} + \delta < \varepsilon.$$

Moreover, for all $y, y' \in Y$, we have

$$|d(h(y), h(y')) - d(y, y')| \leq |d(h(y), h(y')) - d(j(y), j(y'))|$$
$$+ |d(j(y), j(y')) - d(y, y')|$$

$$\leq |d(h(y), h(y')) - d(h(y), j(y'))|$$
$$+ |d(h(y), j(y')) - d(j(y), j(y'))| + \delta$$
$$\leq d(h(y'), j(y')) + d(h(y), j(y)) + \delta$$
$$< 2\bar{\varepsilon} + \delta < \varepsilon,$$

i.e., $h : Y \to X$ is an ε-isometry.

On the other hand, since

$$d(f^n(h(g(y))), j(g^n(g(y)))) \leq \bar{\varepsilon} \quad \text{and} \quad d(f^n(f(h(y))), j(g^n(g(y)))) \leq \bar{\varepsilon}$$

for all $n \in \mathbb{Z}$, we have $f(h(y)) = h(g(y))$ for all $y \in Y$. Therefore, $f \circ h = h \circ g$. It remains to prove that h is continuous.

Fix $\Delta > 0$. Since e is an expansivity constant of f, there is $N \in \mathbb{N}^+$ such that $d(a, b) \leq \Delta$ whenever $d(f^n(a), f^n(b)) \leq e$ for every $-N \leq n \leq N$ (see [76]).

Since g is continuous and Y compact, we have that g is uniformly continuous. So, there is $\gamma > 0$ such that $d(g^n(y), g^n(y')) \leq \frac{\bar{\varepsilon}}{8}$ for all $-N \leq n \leq N$ whenever $y, y' \in Y$ satisfy $d(y, y') < \gamma$.

Then, whenever $d(y, y') < \gamma$, we have

$$d(f^n(h(y)), f^n(h(y'))) = d(h(g^n(y)), h(g^n(y')))$$
$$\leq d(h(g^n(y)), j(g^n(y))) + d(j(g^n(y)), j(g^n(y')))$$
$$+ d(h(g^n(y')), j(g^n(y')))$$
$$\leq 2\bar{\varepsilon} + \delta + d(g^n(y), g^n(y'))$$
$$\leq 3\bar{\varepsilon} + \frac{\bar{\varepsilon}}{8} < e$$

for every $-N \leq n \leq N$. So, $d(h(y), h(y')) \leq \Delta$ by the choice of N. Then h is continuous and the proof is complete.   □

Example 2.4 One can use Exercise 7.4.7 in [14, p. 261] to prove directly from the definition that every homeomorphism of a finite metric space is topologically *GH*-stable (see Exercise 2.14). But this also follows from Theorem 2.3 because all such homeomorphisms are expansive and have the shadowing property.

The next example is on infinite compact metric spaces (including manifolds).

Example 2.5 Let $f : T^2 \to T^2$ be an Anosov diffeomorphism (cf. [3]) of the two-torus T^2. It is well-known that f is expansive and has the shadowing property. Then f is topologically *GH*-stable by Theorem 2.3.

2.6   Gromov-Hausdorff Stability for Group Actions

The results in this section contain part of Dong's dissertation [25]. We extend the Gromov-Hausdorff stability from homeomorphisms to finitely generated group actions. This is the content of the definition below.

Definition 2.6 Let G be a finitely generated group and A be a finite generating set of G. We say that $T \in \text{Act}(G, X)$ is *topologically GH-stable with respect to A* if for every $\varepsilon > 0$, there is $\delta > 0$ such that for every compact metric space Y and every $S \in \text{Act}(G, Y)$ with $d_{\text{GH}^0, A}(T, S) \le \delta$, there is a continuous ε-isometry $h : Y \to X$ such that $T_g \circ h = h \circ S_g$ for every $g \in G$.

We first prove that the above notion of topological GH-stability does not depend on finite generating sets.

Theorem 2.6 *If $T \in \text{Act}(G, X)$ is topologically GH-stable with respect to some finite generating set of G, then T is topologically GH-stable with respect to any finite generating set of G.*

Proof Let A and B be finite generators of G. Suppose that $T \in \text{Act}(G, X)$ is topologically GH-stable with respect to A and take $\varepsilon > 0$. Let $\delta' > 0$ be given by the topological GH-stability of T with respect to A. For this δ', we choose $\delta > 0$ as in Lemma 1.9. Then if $S \in \text{Act}(G, Y)$ for some compact metric space Y and $d_{\text{GH}^0, B}(T, S) < \delta$, then $d_{\text{GH}^0, A}(T, S) < \delta'$ by Lemma 1.9. So, there is a continuous ε-isometry $h : Y \to X$ such that $T_g \circ h = h \circ S_g$ for every $g \in G$. This completes the proof.   □

This theorem leads to the following definition.

Definition 2.7 We say that $T \in \text{Act}(G, X)$ is *topologically GH-stable* if T is topologically *GH*-stable with respect to some finite generating set of G.

Let us present two sufficient conditions for a finitely generated group action of a compact metric space to be topologically GH-stable.

Given $T \in \text{Act}(G, X)$, a finite generating set A, and $\delta > 0$, a *δ-pseudo orbit of T with respect to A* is a sequence $\{x_g\}_{g \in G}$ in X such that $d(T_a(x_g), x_{ag}) < \delta$ for all $a \in A$ and $g \in G$. We say that $\{x_g\}_{g \in G}$ can be *ε-shadowed* if there is $x \in X$ such that $d(T_g(x), x_g) < \varepsilon$ for all $g \in G$.

Definition 2.8 ([57]) We say that T has the *shadowing property* if there is a finite generating set A of G with the following property *(called the shadowing property with respect to A)*: For every $\varepsilon > 0$, there exists $\delta > 0$ such that any δ-pseudo orbit $\{x_g\}_{g \in G}$ with respect to A can be ε-shadowed.

It is well known that this concept does not depend on finite generating sets [63]. The next is the classical definition of expansivity for group actions.

Definition 2.9 We say that $T \in \text{Act}(G, X)$ is *expansive* if there is $c > 0$, called an *expansivity constant* of T, such that if $x, y \in X$ and $d(T_g x, T_g y) \leq c$ for all $g \in G$, then $x = y$.

With these definitions, we can state the following result, generalizing Theorem 2.3.

Theorem 2.7 *If T is an expansive action with the shadowing property of a finitely generated group on a compact metric space, then T is topologically GH-stable.*

Proof Let X be a compact metric space and G be a finitely generated group. Let $T \in \text{Act}(G, X)$ be expansive with the shadowing property. Let η be an expansivity constant of the action T. Fix $\varepsilon > 0$ and take $0 < \bar{\varepsilon} < \frac{1}{8} \min\{\varepsilon, \eta\}$. Let A be a finite generating set of G. Choose δ corresponding to $\bar{\varepsilon}$ as in the definition of shadowing property with respect to A, and assume $\delta < \bar{\varepsilon}$.

Let Y be a compact metric space and $S \in \text{Act}(G, Y)$ with $d_{\text{GH}^0, A}(T, S) < \delta$. Then there are δ-isometries $i : X \to Y$ and $j : Y \to X$ such that

$$\sup_{x \in X, a \in A} d^Y(S_a \circ i(x), i \circ T_a(x)) < \delta \quad \text{and} \quad \sup_{y \in Y, a \in A} d^X(j \circ S_a(y), T_a \circ j(y)) < \delta.$$

Choose $y \in Y$ and consider the sequence $\{x_g\}_{g \in G} \subset X$ defined by $x_g = j \circ S_g(y)$ for all $g \in G$. Since

$$d^X(x_{ag}, T_a(x_g)) = d^X(j \circ S_{ag}(y), T_a \circ (j \circ S_g(y)))$$

$$= d^X((j \circ S_a)S_g(y), (T_a \circ j)S_g(y)) < \delta$$

for all $a \in A$ and $g \in G$, we have that $\{x_g\}_{g \in G} \subset X$ is a δ-pseudo orbit of T with respect to A.

By the shadowing property of T, there is $x \in X$ such that $d(T_g(x), x_g) < \bar{\varepsilon}$ for all $g \in G$. In particular, $d(T_g(x), x_g) < \bar{\varepsilon} < \frac{\eta}{2}$ for all $g \in G$. Since η is an expansivity constant, we see that such an x is unique. Then, by denoting $x = h(y)$, we obtain a map $h : Y \to X$ satisfying

$$d(T_g(h(y)), j(S_g(y))) < \bar{\varepsilon} \tag{2.1}$$

for all $y \in Y$ and $g \in G$. Taking here $g = e$ above, we get $d(h(y), j(y)) < \bar{\varepsilon}$ for all $y \in Y$. Then

$$d_H(h(Y), X) \leq d_H(h(Y), j(Y)) + d_H(j(Y), X) \leq \bar{\varepsilon} + \delta < 2\bar{\varepsilon} < \varepsilon.$$

Moreover, for any $y, y' \in Y$, we have

$$
\begin{aligned}
|d^X(h(y), f(y')) - d^Y(y, y')| &\leq |d^X(h(y), f(y')) - d^X(j(y), j(y'))| \\
&\quad + |d^X(j(y), j(y')) - d^Y(y, y')| \\
&\leq |d^X(h(y), h(y')) - d^X(h(y), j(y'))| \\
&\quad + |d^X(h(y), j(y')) - d^X(j(y), j(y'))| + \delta \\
&\leq d^X(h(y'), j(y')) + d^X(h(y), j(y)) + \delta \\
&< 2\bar{\varepsilon} + \delta < 3\bar{\varepsilon} < \varepsilon,
\end{aligned}
$$

so $h : Y \to X$ is an ε-isometry.

Now, given $a \in A$ and replacing y by $S_a(y)$ in (2.1), we get

$$
d(T_g(h(S_a(y))), j(S_g(S_a(y)))) < \bar{\varepsilon},
$$

and replacing g by ga again in (2.1), we have

$$
d(T_g(T_a(h(y))), j(S_g(S_a(y)))) < \bar{\varepsilon}
$$

for all $g \in G$. Therefore,

$$
\begin{aligned}
d(T_g(T_a(h(y))), T_g(T_a(h(y)))) &\leq d(T_g(h(S_a(y))), j(S_g(S_a(y)))) \\
&\quad + d(T_g(T_a(f(y))), j(S_g(S_a(y)))) \\
&< 2\bar{\varepsilon} < \eta
\end{aligned}
$$

for all $a \in A$ and $g \in G$. Since η is an expansivity constant, $h(S_a(y)) = T_a(h(y))$ for all $y \in Y$ and $a \in A$. Therefore, $T_a \circ h = h \circ S_a$ for all $a \in A$. Since A generates G, we have $T_g \circ h = h \circ S_g$ for all $g \in G$.

It remains to verity that h is continuous. Fix a number $\Delta > 0$. By Lemma 1.10, there exists a finite subset $F \subset G$ such that

$$
\sup_{g \in F} d(T_g(x), T_g(x')) \leq \eta \text{ implies } d^X(x, x') < \Delta.
$$

Since S_g is uniformly continuous for all $g \in F$, there is $\delta_1 > 0$ such that

$$
d^Y(S_g(y), S_g(y')) < \frac{\bar{\varepsilon}}{8}
$$

for all $g \in F$ and $y, y' \in Y$ satisfying $d^Y(y, y') < \delta_1$. Then, for any $y, y' \in Y$ with $d^Y(y, y') < \delta_1$ and $g \in F$, we have

$$d^X(T_g(h(y)), T_g(h(y'))) = d^X(h(S_g(y)), h(S_g(y')))$$
$$\leq d^X(h(S_g(y)), j(S_g(y))) + d^X(j(S_g(y)), j(S_g(y')))$$
$$+ d^X(j(S_g(y')), h(S_g(y')))$$
$$< 2\bar{\varepsilon} + \delta + d^Y(S_g(y), S_g(y'))$$
$$< 3\bar{\varepsilon} + \frac{\bar{\varepsilon}}{8} < \eta.$$

Then $d^X(h(y), h(y')) < \Delta$, and so h is continuous. This completes the proof. $\qquad\square$

This result and the proof of Theorem 2.11 in [23] yield.

Example 2.10 The above theorem implies that the group actions considered in Example 2.14 in [23] are topologically *GH*-stable. In particular, there is a topologically *GH*-stable action of the discrete Heisenberg group on the torus T^{3n} for every $n \geq 2$.

We say that two actions $T \in \text{Act}(G, X)$ and $S \in \text{Act}(G, Y)$ are *isometrically conjugate* if there is an isometry $j : Y \to X$ such that $j \circ S_g = T_g \circ j$ for every $g \in G$. It is known that topological stability is invariant under topological conjugacy. This motivates our last result.

Theorem 2.8 *Let X and X' be compact metric spaces and G be a finitely generated group. If $T \in \text{Act}(G, X)$ and $T' \in \text{Act}(G, X')$ are isometrically conjugate, then T is topologically GH-stable if and only if T' is topologically GH-stable.*

Proof With X, X', T and T' as in the statement of the theorem, and T and T isometrically conjugate, we just need to prove that if T is topologically GH-stable, then so is T'.

Fix a finite generating set A of G and $\varepsilon > 0$, and let $\delta > 0$ be given by the definition of the topological GH-stability of T with respect to A. Since T and T' are isometrically conjugate, $d_{\text{GH}^0, A}(T, T') = 0$ by Item (2) of Lemma 1.7.

Now choose $S' \in \text{Act}(G, Y')$ such that $d_{\text{GH}^0, A}(T', S') < \frac{\delta}{2}$. Then, by Item (1) of Lemma 1.7,

$$d_{\text{GH}^0, A}(T, S') \leq 2(d_{\text{GH}^0, A}(T, T') + d_{\text{GH}^0, A}(T', S')) < \delta.$$

So, by the definition of topological GH-stability of T, there is a continuous ε-isometry $h : Y' \to X$ such that $T_g \circ h = h \circ S'_g$ for every $g \in G$. Since T and T' are isometrically conjugate, there is an isometry $j : X' \to X$ such that $T_g = j \circ T_g' \circ j^{-1}$ for every $g \in G$. Replacing above, we get $j \circ T_g' \circ j^{-1} \circ h = h \circ S_g'$ for all $g \in G$. Define $l = j^{-1} \circ h : Y' \to X'$. Then $j \circ T_g' \circ l = j \circ l \circ S_g'$, so $T_g' \circ l = l \circ S_g'$ for all $g \in G$. Since j is an isometry and h is a continuous ε-isometry, we easily get that l is a continuous ε-isometry. Therefore, T' is topologically GH-stable with respect to A, as desired. $\square$

2.7   Gromov-Hausdorff Stability of Global Attractors

The results of this section will be applied to the Gromov-Hausdorff stability of partial differential equations (Chaps. 5, 6, and 7). Let X be a metric space and $S : X \times \mathbb{R}_0^+ \to X$ be a continuous map. We say that $S(t)$ is a semidynamical system (or a semiflow) on X if $S(x, 0) = x$ and $S(S(x, t_1), t_2) = S(x, t_1 + t_2)$ for $x \in X$ and $t_1, t_2 \geq 0$. For simplicity, we denote $S(\cdot, t)$ by $S(t)$.

Definition 2.11 We say that a compact invariant set $\mathcal{A}$ is the global attractor of $S(t)$ if it attracts all bounded sets, that is, $\mathrm{dist}_X(S(t)B, \mathcal{A}) \to 0$ as $t \to \infty$ for any bounded set $B \subset X$, where

$$\mathrm{dist}_X(A, B) = \sup_{a \in A} \inf_{b \in B} d_X(a, b)$$

for all $A, B \subset X$.

Definition 2.12 Let H be a Hilbert space with a norm $\|\cdot\|$. We say that $\mathcal{M}$ is an inertial manifold if it is a finite-dimensional Lipschitz manifold that is invariant under $S(t)$ and attracts all trajectories exponentially, that is

$$\mathrm{dist}(S(t)u_0, \mathcal{M}) \leq C(\|u_0\|)e^{-kt}$$

for all $u_0 \in H$.

Let Λ be a topological space and $\{S_\lambda : X_\lambda \times \mathbb{R}_0^+ \to X_\lambda\}_{\lambda \in \Lambda}$ be a family of semidynamical systems on metric spaces $\{X_\lambda\}_{\lambda \in \Lambda}$. For each $\lambda_0 \in \Lambda$, let d_λ be a metric on X_λ. Suppose that each system S_λ has a global attractor $\mathcal{A}_\lambda$.

Definition 2.13 We say that the global attractor $\mathcal{A}_\lambda$ ($\lambda \in \Lambda$) of the semi-dynamical system S_λ is Gromov-Hausdorff stable with respect to λ if for each $\lambda_0 \in \Lambda$ and $\varepsilon > 0$, there exists a neighborhood U_{λ_0} of λ_0 such that if $\lambda \in U_{\lambda_0}$, then there are ε-isometries $i : \mathcal{A}_\lambda \to \mathcal{A}_{\lambda_0}$ and $j : \mathcal{A}_{\lambda_0} \to \mathcal{A}_\lambda$, and $\alpha \in \mathrm{Rep}_{\mathcal{A}_\lambda}(\varepsilon)$ and $\beta \in \mathrm{Rep}_{\mathcal{A}_{\lambda_0}}(\varepsilon)$ such that for any $u_\lambda \in \mathcal{A}_\lambda$, $u_{\lambda_0} \in \mathcal{A}_{\lambda_0}$, and $t \in [0, 1]$,

$$d_{\lambda_0}(i(S_\lambda(u_\lambda, \alpha(u_\lambda, t))) - S_{\lambda_0}(i(u_\lambda), t))) < \varepsilon, \text{ and}$$

$$d_\lambda(j(S_{\lambda_0}(u_{\lambda_0}, \beta(u_{\lambda_0}, t))) - S_\lambda(j(u_{\lambda_0}), t))) < \varepsilon.$$

Note that if a semidynamical system $S(t)$ on a metric space X is injective and has the global attractor $\mathcal{A}$, then the restriction of $S(t)$ to $\mathcal{A}$ is a dynamical system on $\mathcal{A}$, which will be denoted by $(\mathcal{A}, \phi)$.

The following theorem gives sufficient conditions for the global attractor $\mathcal{A}_\lambda$ of the semidynamical system S_λ to be GH-stable with respect to $\lambda \in \Lambda$.

Theorem 2.9 *Let Λ be a topological space and $\{S_\lambda : X_\lambda \times \mathbb{R}_0^+ \to X_\lambda\}_{\lambda \in \Lambda}$ be a family of semidynamical systems on metric spaces $\{X_\lambda\}_{\lambda \in \Lambda}$. Suppose that for each $\lambda \in \Lambda$, S_λ has the global attractor $\mathcal{A}_\lambda$ and the restriction on $\mathcal{A}_\lambda$ induces the dynamical systems $(\phi_\lambda, \mathcal{A}_\lambda)$ and satisfies the following:*

(1) *$\mathcal{A}_\lambda$ varies continuously with respect to $\lambda \in \Lambda$ in the Gromov-Hausdorff distance D_{GH^0}, that is, for $\lambda_0 \in \Lambda$ and $\varepsilon > 0$, there exists a neighborhood U_{λ_0} of λ_0 such that for any $\lambda \in U_{\lambda_0}$, $D_{\mathrm{GH}^0}(\mathcal{A}_\lambda, \mathcal{A}_{\lambda_0}) < \varepsilon$;*
(2) *for any $\lambda_0 \in \Lambda$, there is a neighborhood U_{λ_0} of λ_0 such that $\{S_\lambda\}_{\lambda \in U_{\lambda_0}}$ is equicontinuous on $\mathcal{A}_\lambda \times [0, 1]$, that is, for any $\varepsilon > 0$, there exists $\delta > 0$ such that if $d_{X_\lambda}(x, y) < \delta$ and $|t - s| < \delta$, then $d_{X_\lambda}(S_\lambda(x, t), S_\lambda(y, s)) < \varepsilon$ for any $\lambda \in U_{\lambda_0}$, $x, y \in \mathcal{A}_\lambda$, and $t, s \in [0, 1]$;*
(3) *for the ε-isometries $i : \mathcal{A}_\lambda \to \mathcal{A}_{\lambda_0}$ and $j : \mathcal{A}_{\lambda_0} \to \mathcal{A}_\lambda$ obtained in (1),*

$$d_{C^0}(j \circ i, id_{\mathcal{A}_\lambda}), \ d_{C^0}(i \circ j, id_{\mathcal{A}_{\lambda_0}}) < \varepsilon, \text{ and}$$

$$d_{\lambda_0}(i S_\lambda j(x, t), S_{\lambda_0}(x, t)) < \varepsilon$$

uniformly for $(x, t) \in \mathcal{A}_{\lambda_0} \times [0, 1]$ and $\lambda \in U_{\lambda_0}$.

Then the global attractor $\mathcal{A}_\lambda$ of the semidynamical system S_λ is GH-stable with respect to $\lambda \in \Lambda$.

Proof For any $\varepsilon > 0$, by the assumption (2), we can choose a constant $0 < \delta < \dfrac{\varepsilon}{3}$ such that if $u, \tilde{u} \in \mathcal{A}_\lambda$ and $t, s \in [0, 1]$ with $d_\lambda(u, \tilde{u}) < \delta$ and $|t - s| < \delta$, then

$$d_\lambda(\phi_\lambda(u, t), \phi_\lambda(\tilde{u}, s)) < \varepsilon.$$

For $\lambda_0 \in \Lambda$, by the assumptions (1) and (3), we see that there is a neighborhood U_{λ_0} of λ_0 such that for any $\lambda \in U_{\lambda_0}$, there exist δ-isometries $\hat{i} : \mathcal{A}_\lambda \to \mathcal{A}_{\lambda_0}$ and $\hat{j} : \mathcal{A}_{\lambda_0} \to \mathcal{A}_\lambda$ such that

$$d_{C^0}(j \circ i, id_{A_\lambda}), \ d_{C^0}(i \circ j, id_{A_{\lambda_0}}) < \delta \text{ and } d_{\lambda_0}(\hat{i}(\phi_\lambda \hat{j}(u, t)), \phi_{\lambda_0}(u, t)) < \delta$$

for every $(u, t) \in \mathcal{A}_{\lambda 0} \times [0, 1]$. Then we have

$$d_\lambda(\phi_\lambda(\hat{j}(u), t), \hat{j}(\phi_{\lambda_0}(u, t)))$$

$$\leq d_{\lambda_0}(\hat{i}(\phi_\lambda(\hat{j}(u), t)), \hat{i} \circ \hat{j}(\phi_{\lambda_0}(u, t))) + \frac{\varepsilon}{3}$$

$$\leq d_\lambda(\hat{i}(\phi_\lambda(\hat{j}(u), t)), \phi_{\lambda_0}(u, t)) + d_{\lambda_0}(\phi_{\lambda_0}(u, t), \hat{i} \circ \hat{j}(\phi_{\lambda_0}(u, t))) + \frac{\varepsilon}{3} < \varepsilon.$$

Moreover,

$$d_{\lambda_0}(\hat{i}(\phi_\lambda(u, t)), \phi_{\lambda_0}(\hat{i}(u), t))$$

$$\leq d_{\lambda_0}(\hat{i}(\phi_\lambda(u, t)), \hat{i}(\phi_\lambda(\hat{j} \circ \hat{i}(u), t))) + d_{\lambda_0}(\hat{i}(\phi_\lambda(\hat{j} \circ \hat{i}(u), t)), \phi_{\lambda_0}(\hat{i}(u), t))$$

$$\leq d_\lambda(\phi_\lambda(u, t), \phi_\lambda(\hat{j} \circ \hat{i}(u), t)) + \frac{2\varepsilon}{3} < \varepsilon.$$

This shows that

$$D_{GH^0}(\phi_\lambda, \phi_{\lambda_0}) < \varepsilon$$

for all $\lambda \in U_{\lambda_0}$, and completes the proof of the theorem.   □

Remark 2.1 We can derive that if X, Y are compact metric spaces and $i : X \to Y$ is an ε-isometry, then there exists a 5ε-isometry $j : Y \to X$ such that

$$d_{C^0}(j \circ i, id_X), \ d_{C^0}(i \circ j, id_Y) < 5\varepsilon.$$

   We will need the following proposition, which can be considered as a generalization of Theorem 5.2 in [37].

Proposition 2.1 *Let Λ be a topological space and $\{S_\lambda : X_\lambda \times \mathbb{R}_0^+ \to X_\lambda\}_{\lambda \in \Lambda}$ be a family of semidynamical systems on metric spaces $\{X_\lambda\}_{\lambda \in \Lambda}$. Suppose that for each $\lambda \in \Lambda$,*

(1) *S_λ has the global attractor $\mathcal{A}_\lambda$;*
(2) *there is a bounded open neighborhood D_λ of $\mathcal{A}_\lambda$ such that for any $t \in \mathbb{R}_0^+$, $\lambda_0 \in \Lambda$, and $\varepsilon > 0$, there exists a neighborhood W of λ_0 such that for any $\lambda \in W$, there is an ε-immersion $i_\lambda : D_\lambda \to X_{\lambda_0}$ such that*
   (i) *$i_\lambda(D_\lambda) \subset B(D_{\lambda_0}, \varepsilon)$ and*
   (ii) *$d(i_\lambda(S_\lambda(x, t)), S_{\lambda_0}(i_\lambda(x), t)) < \varepsilon$ for all $x \in \mathcal{A}_\lambda$.*

Then, the map $\mathbb{A} : \Lambda \to \mathcal{M}$ *given by* $\mathbb{A}(\lambda) = \mathcal{A}_\lambda$ *is residually continuous.*

Proof Step 1. We first show that for any $\lambda_0 \in \Lambda$ and $\varepsilon > 0$, there exists a neighborhood W of λ_0 such that for any $\lambda \in W$, there is an ε-immersion $i_\lambda : D_\lambda \to X_{\lambda_0}$ such that $i_\lambda(B(\mathcal{A}_\lambda, \varepsilon/4) \cap D_\lambda) \subset B(\mathcal{A}_{\lambda_0}, \varepsilon)$.

Since $\mathcal{A}_{\lambda_0}$ is the global attractor of $S_{\lambda_0}(t)$, there is $T > 0$ such that

$$S_{\lambda_0}(B(D_{\lambda_0}, \varepsilon), T) \subset B(\mathcal{A}_{\lambda_0}, \varepsilon/4).$$

By the assumption (2), there exists a neighborhood W of λ_0 such that for $\lambda \in W$, there exists an $\varepsilon/4$-immersion $i_\lambda : D_\lambda \to X_{\lambda_0}$ such that

$$i_\lambda(D_\lambda) \subset B(D_{\lambda_0}, \varepsilon/4) \text{ and } d(i_\lambda(S_\lambda(x, T)), S_{\lambda_0}(i_\lambda(x), T)) < \frac{\varepsilon}{4}$$

for any $x \in \mathcal{A}_\lambda$. Since $\mathcal{A}_\lambda$ is invariant, for any $y \in B(\mathcal{A}_\lambda, \varepsilon/4) \cap D_\lambda$, there is $x \in \mathcal{A}_\lambda$ such that $d(y, S_\lambda(x, T)) < \varepsilon/4$. Consequently,

$$d(i_\lambda(y), \mathcal{A}_{\lambda_0}) \le d(i_\lambda(y), i_\lambda(S_\lambda(x, T))) + d(i_\lambda(S_\lambda(x, T)), S_{\lambda_0}(i_\lambda(x), T))$$

$$+ d(S_{\lambda_0}(i_\lambda(x), T), \mathcal{A}_{\lambda_0})$$

$$\le d(y, S_\lambda(x, T)) + \frac{\varepsilon}{4} + \frac{\varepsilon}{4} + \frac{\varepsilon}{4} < \varepsilon.$$

Hence, we get $i_\lambda(B(\mathcal{A}_\lambda, \varepsilon/4) \cap D_\lambda) \subset B(\mathcal{A}_{\lambda_0}, \varepsilon)$, and the proof of Step 1 is complete.

Step 2. For any $\varepsilon > 0$, we denote by $E(\varepsilon)$ the collection of $\lambda_0 \in \Lambda$ such that there is a neighborhood W of λ_0 with the following property: for any $\lambda, \lambda' \in W$, $D_{\mathrm{GH}^0}(\mathcal{A}_\lambda, \mathcal{A}_{\lambda'}) < \varepsilon$. We observe that $\bigcap_{n \in \mathbb{N}} E(1/n)$ consists of all points of continuity of the map $\mathbb{A} : \Lambda \to \mathcal{M}$.

To prove that the map $\mathbb{A}$ is residually continuous, we show that $E(\varepsilon)$ is open and dense in Λ for any $\varepsilon > 0$. It is clear that $E(\varepsilon)$ is open in Λ, and so it is enough to show that $E(4\varepsilon)$ is dense in Λ for any $\varepsilon > 0$. Let $\varepsilon > 0$ and let U be a nonempty open set in Λ. Fix $\lambda_0 \in U$, and take a finite open cover $\{B_i\}_{i \in K}$ of $\mathcal{A}_{\lambda_0}$ in X_{λ_0} such that

$$\mathrm{diam}(B_i) < \varepsilon \text{ and } \mathcal{A}_{\lambda_0} \cap B_i \ne \emptyset$$

for all $i \in K$. Let $\mathcal{I}$ be the collection of all subsets J of K such that there are $\lambda_1 \in U$ and a δ-immersion $i_{\lambda_1} : D_{\lambda_1} \to X_{\lambda_0}$ for some $\delta < \varepsilon/2$ satisfying

$$i_{\lambda_1}(\mathcal{A}_{\lambda_1}) \cap B_i \ne \emptyset \text{ for all } i \in J, \text{ and}$$

$$i_{\lambda_1}(U_{\lambda_1}) \subset \bigcup_{i \in J} B_i \text{ for an open neighborhood } U_{\lambda_1} \text{ of } \mathcal{A}_{\lambda_1}.$$

It is clear that $\mathcal{I}$ is nonempty since $K \in \mathcal{I}$. Since K is finite, we can choose a minimal element J of $\mathcal{I}$. Take $\lambda_1 \in U$ and a δ-immersion i_{λ_1} corresponding to J by the definition of $\mathcal{I}$. Let $W \subset U$ be a neighborhood of λ_1 chosen as in Step 1. For any $\lambda \in W$, there is an $(\varepsilon/4 - \delta/2)$-immersion $i_\lambda : D_\lambda \to X_{\lambda_1}$ such that $i_\lambda(U_\lambda) \subset U_{\lambda_1}$ for a neighborhood U_λ of $\mathcal{A}_\lambda$. It is clear that $i_{\lambda_1} \circ i_\lambda$ is an $(\varepsilon/4 + \delta/2)$-immersion from D_λ to X_{λ_1} such that

$$ i_{\lambda_1} \circ i_\lambda(U_\lambda) \subset i_{\lambda_1}(U_{\lambda_1}) \subset \bigcup_{i \in J} B_i. $$

By the minimality of J, we have $i_{\lambda_1} \circ i_\lambda(\mathcal{A}_\lambda) \cap B_i \neq \varnothing$ for all $i \in J$. This implies that $D_{GH^0}(\mathcal{A}_\lambda, \mathcal{A}_{\lambda_1}) < 2\varepsilon$ for any $\lambda \in W$, and so $\lambda_1 \in E(4\varepsilon)$. This completes the proof of the proposition. □

Exercises

Exercise 2.14 Use Exercise 7.4.7 in [14, p. 261] to deduce directly from the definition that every homeomorphism of a finite metric space is topologically GH-stable.

Recall that $x \in X$ is a *periodic point* of a map $f : X \to X$ if there is a positive integer n such that $f^n(x) = x$.

Exercise 2.15 Prove that if a transitive homeomorphism of a compact metric space $f : X \to X$ is topologically GH-stable, then the periodic orbits of f are dense in X.

Exercise 2.16 Is every topologically stable circle homeomorphism topologically GH-stable?

Recall that a continuous map $f : X \to X$ of a compact metric space is *positively expansive* if there is $\varepsilon > 0$ such that if $x, x' \in X$ and $d(f^n(x), f^n(x')) \leq \varepsilon$ for all $n \in \mathbb{N}$, then $x = y$. Moreover, f has the *shadowing property* if for every $\varepsilon > 0$, there is $\delta > 0$ such that for all sequence $\{x_n\}_{n \in \mathbb{N}}$ with $d(f(x_n), x_{n+1}) \leq \delta$ for all $n \geq \mathbb{N}$, there is $x \in X$ such that $d(f^n(x), x_n) \leq \varepsilon$ for all $n \in \mathbb{N}$.

Exercise 2.17 Taking into account the remark after Definition 5.2, prove that every positively expansive continuous map with the shadowing property of a compact metric space is topologically GH-stable.

Given a map $f : X \to X$ of a metric space, a δ-*chain* is a finite set $x_0, \ldots, x_k$ such that $d(f(x_i), x_{i+1}) < \delta$ for all $0 \leq i \leq k - 1$. A δ-chain from x to y is a δ-chain with $x_0 = x$ and $x_k = y$. We write $x \sim y$ if for any $\delta > 0$ there are δ-chains from x to y and from y to

x. Define $\mathcal{CR}(f)$, the *chain recurrent set*, of f, by the rule: as $x \in \mathcal{CR}(f)$ if and only if $x \sim x$. Recall that $\mathrm{Per}(f)$ denotes the set of periodic points of f.

Exercise 2.18 Is $\mathcal{CR}(f) = \overline{\mathrm{Per}(f)}$ true for every topologically GH-stable homeomorphism of a compact metric space $f : X \to X$?

Continuity of the Shift Operator

3

3.1 Introduction

The relationships between continuous maps $f : X \to X$ and their corresponding limit inverse representation $\sigma_f : X_f \to X_f$ have been studied in the literature. For example, Chen and Li [17] proved that f have the shadowing or asymptotically shadowing properties if and only if σ_f does. In [12], the author showed equivalences between the ergodic shadowing property of f and of σ_f. Several other properties were investigated in [43]. More recently, Tsegmid [74] proved that if f has the average shadowing property, then so does σ_f. The stability properties of f and those of σ_f have also been investigated. In [78], the orbit shift structural stability of hyperbolic self covering was obtained. Sun [71] defined orbit shift topological stability and proved that this property holds for all surjective Anosov maps (see also [69]). Related results are given in [19] and [18].

In this chapter, we will investigate how $\sigma_f : X_f \to X_f$ varies with respect to Gromov-Hausdorff perturbations of $f : X \to X$. Indeed, we show that the map $f \mapsto \sigma_f$ (called the *shift operator*) is continuous (with respect to such perturbations) at every topologically Anosov map f. We apply this result to the stability theory of topological dynamical systems. Let us present these results in detail.

Let C_{sur} denote the set of surjective continuous maps of compact metric spaces $f : X \to X$. A special subset of C_{sur} is the set of homeomorphisms $f : X \to X$, denoted by H. Let $\mathcal{C}_{\text{sur}}$ denote the corresponding set of equivalence classes $[f]$ for $f \in C_{\text{sur}}$ and, correspondingly, let $\mathcal{H} \subset \mathcal{C}_{\text{sur}}$ denote the set of equivalence classes corresponding to elements $f \in H$. Then d_{GH^0} induces a quasi-metric (and thus a topology) on $\mathcal{C}_{\text{sur}}$. All continuity properties for maps defined in $\mathcal{C}_{\text{sur}}$ will refer to this topology.

© The Author(s), under exclusive license to Springer Nature Switzerland AG 2022
J. Lee, C. Morales, *Gromov-Hausdorff Stability of Dynamical Systems and Applications to PDEs*, Frontiers in Mathematics, https://doi.org/10.1007/978-3-031-12031-2_3

Let us define the shift operator $\sigma : C_{\text{sur}} \to \mathcal{H}$. Let X be a compact metric space. Denote by $X^{\mathbb{Z}}$ the set of bi-sequences $\xi = \{\xi_i\}_{i \in \mathbb{Z}}$ with $\xi_i \in X$ for every $i \in \mathbb{Z}$. It follows that $X^{\mathbb{Z}}$ is compact if endowed with the metric

$$D(\xi, \bar{\xi}) = \sum_{i \in \mathbb{Z}} 2^{-|i|} d(\xi_i, \bar{\xi}_i).$$

Any surjective map $f : X \to X$ induces a map $\tilde{f} : X^{\mathbb{Z}} \to X^{\mathbb{Z}}$ given by $\tilde{f}(\xi) = (f(\xi_i))_{i \in \mathbb{Z}}$. We will see in Remark 3.1 that the map $T : C_{\text{sur}} \to C$ defined by $T([f]) = [\tilde{f}]$ is well-defined and continuous with respect to the topology induced by the Gromov-Hausdorff distance.

Now define the inverse limit space of f by

$$X_f = \{\xi \in X^{\mathbb{Z}} : f(\xi_i) = \xi_{i+1} \text{ for all } i \in \mathbb{Z}\}.$$

It follows that X_f is a compact subset of $X^{\mathbb{Z}}$ and also invariant, that is, $\tilde{f}(X_f) = X_f$. The restriction $\sigma_f = \tilde{f}|_{X_f} : X_f \to X_f$ is called the *shift map* associated to f. Notice that if f is continuous, then $\sigma_f : X_f \to X_f$ is a homeomorphism [3]. In particular, $\sigma_f \in H$ for all $f \in C_{\text{sur}}$.

We will see in the next section that the map $\sigma : C_{\text{sur}} \to \mathcal{H}$ defined by

$$\sigma([f]) = [\sigma_f]$$

for all $[f] \in C_{\text{sur}}$ is well-defined. We call this map the *shift operator*. In the sequel, we give sufficient conditions for its continuity at $[f] \in C_{\text{sur}}$. For simplicity, we write f instead of $[f]$.

A continuous surjective map $f : X \to X$ of a compact metric space is c-*expansive* if there is $\delta > 0$ such that if $\xi, \xi' \in X_f$ and $d(\xi_i, \xi'_i) \leq \delta$ for every $i \in \mathbb{Z}$, then $\xi = \xi'$. Given $\delta > 0$, a δ-*pseudo orbit* of f is a sequence $\{x_i\}_{i=1}^{\infty}$ such that $d(f(x_i), x_{i+1}) \leq \delta$ for every non-negative integer i. If $\varepsilon > 0$, then we say that $(x_i)_{i=0}^{\infty}$ can be ε-*shadowed* if there is $x \in X$ such that $d(f^i(x), x_i) \leq \varepsilon$ for every non-negative integer i. We say that f has the *shadowing property* if for every $\varepsilon > 0$, there is $\delta > 0$ such that every δ-pseudo orbit can be ε-shadowed.

Definition 3.1 ([3]) A surjective map $f : X \to X$ is *topologically Anosov* if it is c-expansive and has the shadowing property.

It is well known that the property of being topologically Anosov is invariant under topological (and so under isometric) conjugacy [3]. Consequently the property of being topologically Anosov is well-defined in C_{sur}.

Our main result is the following.

Theorem 3.1 *The shift operator is continuous at every topologically Anosov map of a compact metric space with positive diameter.*

The rest of the chapter is organized as follows. In Sect. 3.2, we give some preliminaries. In Sect. 3.3, we prove Theorem 3.1. In Sect. 3.4, we apply this theorem to the stability theory of dynamical systems.

3.2 Preliminary Facts

To start, we prove that the shift operator is well defined.

Lemma 3.1 *If $f, g \in C_{\mathrm{sur}}$ are isometrically conjugate, then so are σ_f and σ_g.*

Proof Let $h : Y \to X$ be an isometry such that

$$f \circ h = h \circ g.$$

Define $H : Y^{\mathbb{Z}} \to X^{\mathbb{Z}}$ by

$$H(\xi)_i = h(\xi_i)$$

for $\xi \in Y^{\mathbb{Z}}$ and $i \in \mathbb{Z}$. Since h is an isometry,

$$D(H(\xi), H(\xi')) = \sum_{i \in \mathbb{Z}} 2^{-|i|} d(h(\xi_i), h(\xi_i')) = \sum_{i \in \mathbb{Z}} d(\xi_i, \xi_i') = D(\xi, \xi'),$$

for all $\xi, \xi' \in Y^{\mathbb{Z}}$. Then H is an isometric immersion (hence, continuous).

On the other hand, if $\xi \in Y_g$, that is, $g(\xi_i) = \xi_{i+1}$ for $i \in \mathbb{Z}$, then

$$fh(\xi_i) = hg(\xi_i) = h(\xi_{i+1}).$$

Thus, $H(\xi) \in X_f$, yielding an isometric immersion $H : Y_g \to X_f$.

Given $\eta \in X_f$, we can define $\xi \in Y^{\mathbb{Z}}$ by

$$\xi_i = h^{-1}(\eta_i)$$

for $i \in \mathbb{Z}$. Clearly, $H(\xi) = \eta$. Since

$$g(\xi_i) = gh^{-1}(\eta_i) = h^{-1} f(\eta_i) = h^{-1}(\eta_{i+1}),$$

we have $\xi \in Y_g$, proving that H is onto (hence, an isometry from Y_g to X_f).

Finally, for every $\xi \in Y_g$ and $i \in \mathbb{Z}$, we have

$$\sigma_f(H(\xi))_i = h(\xi_{i+1}) = H(\sigma_g(\xi))_i,$$

proving $\sigma_f \circ H = H \circ \sigma_g$. This completes the proof.   □

Part of the above proof can be used to prove the following remark.

Remark 3.1 The map $T : \mathcal{C}_{\text{sur}} \to \mathcal{C}$ defined by $T(f) = \tilde{f}$ is both well defined and continuous with respect to the topology induced by the Gromov-Hausdorff distance.

Proof We can see as above that if $f, g \in \mathcal{C}$ are isometrically conjugate, then so are $\tilde{f}$ and $\tilde{g}$. This implies that T is well defined.

Now fix ε. For a suitable $\delta > 0$, we assume that $f, g \in \mathcal{C}_{\text{sur}}$ satisfy $d_{\text{GH}^0}(f, g) < \delta$. Then there are δ-isometries $i : X \to Y$ and $j : Y \to X$ such that

$$d(g \circ i, i \circ f) < \delta \quad \text{and} \quad d(j \circ g, f \circ j) < \delta.$$

Define the maps $I : X^{\mathbb{Z}} \to Y^{\mathbb{Z}}$ and $J : Y^{\mathbb{Z}} \to X^{\mathbb{Z}}$ by

$$I(\xi)_k = i(\xi_k) \quad \text{and} \quad J(\eta)_k = j(\eta_k)$$

for $(k, \xi, \eta) \in \mathbb{Z} \times X^{\mathbb{Z}} \times Y^{\mathbb{Z}}$. Following the proof of the lemma above, we can see that for δ small, the maps I and J are ε-isometries satisfying

$$D(\tilde{g} \circ I, I \circ \tilde{f}) < \varepsilon \quad \text{and} \quad D(J \circ \tilde{g}, \tilde{f} \circ J) < \varepsilon.$$

Hence, $(\tilde{f}, \tilde{g}) < \varepsilon$ and the proof follows.   □

Remark 3.2 The above remark suggests at once that the shift operator $\sigma : \mathcal{C}_{\text{sur}} \to \mathcal{H}$ at every $f \in \mathcal{C}_{\text{sur}}$ should be continuous (this obviously implies Theorem 3.1). However, the proof above breaks in particular when trying to prove the inclusions $I(X_f) \subset Y_g$ and $J(Y_g) \subset X_f$. We define δ-isometries in a different manner when f is topologically Anosov (see Eq. (3.1) in the proof of Theorem 3.1).

The following lemma permits a reduction of the topology induced by d_{GH^0} in $\mathcal{C}$.

Lemma 3.2 *Let* $f : X \to X$ *be a continuous map of a compact metric space. Then for every* $\Delta > 0$, *there exists* $\Delta' > 0$ *such that if* $g : Y \to Y$ *is a continuous map of a compact metric space satisfying* $d(j \circ g, f \circ j) \leq \Delta'$ *for some* Δ'-*isometry* $j : Y \to X$, *then* $d_{\text{GH}^0}(f, g) \leq \Delta$.

Proof Fix $\Delta > 0$. Take $\beta > 0$ such that $d(a, b) \leq \beta$ for $a, b \in X$ implies $d(f(a), f(b)) \leq \frac{\Delta}{4}$. Now fix $0 < \Delta' < \min\{\frac{\Delta}{4}, \beta\}$.

Let $g : Y \to Y$ be a continuous map of a compact metric space such that $d(j \circ g, f \circ j) \leq \Delta'$ for some Δ'-isometry $j : Y \to X$. Then, by Lemma 1.6, there is a $3\Delta'$-isometry $i : X \to Y$ such that

$$d(j \circ i(x), x) < \Delta',$$

for all $x \in X$. Since $\Delta' < \Delta$, i is a Δ-isometry. Then, since

$$d(g \circ i(x), i \circ f(x)) < \Delta' + d(j \circ g \circ i(x), j \circ i \circ f(x))$$
$$\leq \Delta' + d((j \circ g)(i(x)), (f \circ j)(i(x)))$$
$$+ d(f \circ j \circ i(x), j \circ i \circ f(x))$$
$$< 2\Delta' + d(f(j \circ i(x)), f(x)) + d(f(x), j \circ i(f(x)))$$
$$< 3\Delta' + d(f(j \circ i(x)), f(x))$$

and $d(j \circ i(x), x) < \Delta' < \beta$ for every $x \in X$, we get

$$d(g \circ i(x), i \circ f(x)) \leq 3\Delta' + \frac{\Delta}{4} < \frac{3\Delta}{4} + \frac{\Delta}{4} = \Delta$$

for all $x \in X$, proving $d(g \circ i, i \circ f) \leq \Delta$. Again, since $\Delta' < \Delta$, j is a Δ-isometry and $d(j \circ g, f \circ j) \leq \Delta$. Then $d_{\mathrm{GH}^0}(f, g) \leq \Delta$, completing the proof. □

The last lemma is Theorem 2.3.7 in [3, p. 81], restated here for the reader's convenience.

Lemma 3.3 *A continuous surjection* $f : X \to X$ *of a compact metric space has the shadowing property if and only if for every* $\varepsilon' > 0$, *there is* $\delta > 0$ *such that for every* $\hat{\xi} \in X^{\mathbb{Z}}$ *with* $d(\hat{\xi}_i, \hat{\xi}_{i+1}) < \delta$ *(for* $i \in \mathbb{Z}$*), there is* $\xi \in X_f$ *such that* $d(\xi_i, \hat{\xi}_i) < \varepsilon'$ *for every* $i \in \mathbb{Z}$.

3.3　Proof of Theorem 3.1

In this section, we will prove Theorem 3.1. We have to show that if $f : X \to X$ is a surjective c-expansive map with the shadowing property of a compact metric space with diameter $\mathrm{diam}(X) > 0$, then for every $\Delta > 0$, there is $\delta > 0$ such that if a surjective continuous map $g : Y \to Y$ of a compact metric space satisfies $d_{\mathrm{GH}^0}(f, g) \leq \delta$, then $d_{\mathrm{GH}^0}(\sigma_f, \sigma_g) \leq \Delta$.

By Lemma 3.2, it suffices to show that for every $\Delta' > 0$, there is $\delta > 0$ such that if $d_{\mathrm{GH}^0}(f, g) \leq \delta$, then there is a Δ'-isometry $J : Y_g \to X_f$ such that $D(\sigma_f \circ J, J \circ \sigma_g) \leq \Delta'$. We proceed as follows.

Let e be a c-expansivity constant of f, fix $\Delta' > 0$, and take

$$0 < \varepsilon' < \frac{1}{6} \min\{\Delta', e\}.$$

Also, fix a positive integer N such that

$$\sum_{|i| \geq N+1} 2^{-|i|} \leq \frac{\Delta'}{4 \operatorname{diam}(X)}$$

and

$$0 < R < \frac{\Delta'}{12(2N+1)}.$$

Since f is surjective with the shadowing property and $6\varepsilon' < \Delta'$, we can choose

$$0 < \delta < \frac{1}{3}(\Delta' - 6\varepsilon')$$

satisfying the conclusion of Lemma 3.3 for the given ε'. Making δ smaller if necessary, we can assume that

$$d(a, b) \leq \delta \text{ for } a, b \in X \quad \text{implies} \quad d(f^k(a), f^k(b)) \leq R \text{ for } 0 \leq k \leq 2N.$$

Let us verity that this δ works.

Indeed, suppose that $g : Y \to Y$ is a surjective continuous map of a compact metric space such that

$$d_{\mathrm{GH}^0}(f, g) \leq \delta.$$

Then there is a δ-isometry $j : Y \to X$ such that $d(f \circ j, j \circ g) \leq \delta$. Given $\eta \in Y_g$, define $\hat{\xi} \in X^{\mathbb{Z}}$ by

$$\hat{\xi}_i = j(\eta_i)$$

for all $i \in \mathbb{Z}$. Since

$$d(f(\hat{\xi}_i), \hat{\xi}_{i+1}) = d((f \circ j)\eta_i, (j \circ g)\eta_i) \leq d(f \circ j, j \circ g) \leq \delta,$$

Lemma 3.3 allows us to choose $\xi \in X_f$ such that

$$d(\hat{\xi}_i, \xi_i) \leq \varepsilon'.$$

Such a ξ is unique because assumably that there exist another one, $\xi' \in X_f$, we have

$$d(\xi_i, \xi'_i) \leq d(\hat{\xi}_i, \xi_i) + d(\hat{\xi}_i, \xi'_i) \leq 2\varepsilon' \leq e.$$

Hence, $\xi = \xi'$ by c-expansivity. Thus, we have proved that for a given $\eta \in Y_g$, there is a unique $\xi \in X_f$ satisfying

$$d(j(\eta_i), \xi_i) \leq \varepsilon'$$

for all $i \in \mathbb{Z}$. By setting $J(\eta) = \xi$, where ξ is such a unique element, we get a map $J : Y_g \to X_f$ such that

$$d(j(\eta_i), J(\eta)_i) \leq \varepsilon' \tag{3.1}$$

for $\eta \in Y_g$ and $i \in \mathbb{Z}$. Let us prove that the map enjoys the required properties, that is,

$$J \text{ is a } \Delta'\text{-isometry} \quad \text{and} \quad D(\sigma_f \circ J, J \circ \sigma_g) \leq \Delta'.$$

To prove the first property, if $\eta, \eta' \in Y_g$, we observe that

$$\begin{aligned}
D(J(\eta), J(\eta')) &= \sum_{i \in \mathbb{Z}} 2^{-|i|} d(J(\eta)_i, J(\eta')_i) \\
&\leq \sum_{i \in \mathbb{Z}} 2^{-|i|} [d(J(\eta)_i, j(\eta_i)) + d(j(\eta_i), j(\eta'_i)) + d(J(\eta')_i, j(\eta'_i))] \\
&\leq 6\varepsilon' + \sum_{i \in \mathbb{Z}} 2^{-|i|} d(j(\eta_i), j(\eta'_i)) \\
&\leq (6\varepsilon' + 3\delta) + \sum_{i \in \mathbb{Z}} 2^{-|i|} d(\eta_i, \eta'_i) \\
&\leq \Delta' + D(\eta, \eta'),
\end{aligned}$$

and so

$$D(J(\eta), J(\eta')) - D(\eta, \eta') \leq \Delta'$$

for $\eta, \eta' \in Y_g$. Likewise,

$$D(\eta, \eta') = \sum_{i \in \mathbb{Z}} 2^{-|i|} d(\eta_i, \eta'_i)$$

$$\le 3\delta + \sum_{i \in \mathbb{Z}} 2^{-|i|} d(j(\eta_i), j(\eta'_i))$$

$$\le 6\delta + \sum_{i \in \mathbb{Z}} 2^{-|i|} [d(j(\eta_i), J(\eta)_i) + d(J(\eta)_i, J(\eta')_i) + d(j(\eta'_i), J(\eta')_i)]$$

$$\le (6\varepsilon' + 3\delta) + \sum_{i \in \mathbb{Z}} 2^{-|i|} d(J(\eta)_i, J(\eta')_i)$$

$$\le \Delta' + D(J(\eta), J(\eta')),$$

and so $D(\eta, \eta') - D(J(\eta), J(\eta')) \le \Delta'$ for $\eta, \eta' \in Y_g$. Therefore,

$$\sup_{\eta, \eta' \in Y_g} |D(J(\eta), J(\eta')) - D(\eta, \eta')| \le \Delta'. \tag{3.2}$$

To prove the second property, we take $\xi \in X_f$. Since $j : Y \to X$ is a δ-isometry, there is $\eta_{-N} \in Y$ such that $d(j(\eta_{-N}), \xi_{-N}) \le \delta$. Define $\eta_{-N+k} = g^k(\eta_{-N})$ for $k \ge 0$. Since g is surjective, we can extend η_{-N+k} for $k < 0$ to obtain $\eta \in Y_g$. Since

$$d(\xi_{-N+k}, j(\eta_{-N+k})) = d(\xi_{-N+k}, jg^k(\eta_{-N}))$$

and

$$d(\xi_{-N+k}, jg^k(\eta_{-N})) \le d(f^k(\xi_{-N}), f^k(j(\eta_{-N})))$$

$$+ d(f^{k-1}(f \circ j(\eta_{-N})), f^{k-1}(j \circ g(\eta_{-N})))$$

$$+ d(f^{k-2}(f \circ j(g(\eta_{-N}))), f^{k-2}(j \circ g(g(\eta_{-N}))))$$

$$+ d(f^{k-3}(f \circ j(g^2(\eta_{-N}))), f^{k-3}(j \circ g(g^2(\eta_{-N}))))$$

$$\vdots$$

$$+ d(f \circ j(g^{k-1}(\eta_{-N})), j \circ g(g^{k-1}(\eta_{-N})))$$

$$\le (k+1)R \le (2N+1)R \le \frac{\Delta'}{12}$$

for any $0 \le k \le 2N$, we get

$$d(\xi_{-N+k}, j(\eta_{-N+k})) \le \frac{\Delta'}{12}.$$

Consequently,

$$D(J(\eta), \xi) = \sum_{|i| \geq N+1} 2^{-|i|} d(J(\eta)_i, \xi_i) + \sum_{i=-N}^{N} 2^{-|i|} d(J(\eta)_i, \xi_i)$$

$$\leq \operatorname{diam}(X) \left(\sum_{|i| \geq N+1} 2^{-|i|} \right) + \sum_{i=-N}^{N} 2^{-|i|} [d(j(\eta_i), J(\eta)_i) + d(j(\eta_i), \xi_i)]$$

$$\leq \frac{\Delta'}{4} + 3\varepsilon' + \sum_{k=0}^{2N} 2^{(-N+k)} d(\xi_{-N+k}, j(\eta_{-N+k}))$$

$$\leq \frac{\Delta'}{4} + \frac{\Delta'}{2} + \frac{\Delta'}{12} \cdot 3 = \Delta'.$$

It follows that for every $\xi \in X_f$, there is $\eta \in Y_g$ such that $D(J(\eta), \xi) \leq \Delta'$. Therefore,

$$D_{\mathrm{H}}(J(Y_g), X_f) \leq \Delta', \tag{3.3}$$

which together with (3.2) implies that J is a Δ'-isometry.

For the last assertion, fix $\eta \in Y_g$. Then

$$(\sigma_f \circ J(\eta))_i = J(\eta)_{i+1} \quad \text{and} \quad j(\eta_{i+1}) = j(\sigma_g(\eta)_i)$$

for $i \in \mathbb{Z}$. So,

$$d((\sigma_f \circ J(\eta))_i, (J \circ \sigma_g(\eta))_i) = d(J(\eta)_{i+1}, J(\sigma_g(\eta))_i)$$

$$\leq d(J(\eta)_{i+1}, j(\eta_{i+1})) + d(j(\sigma_g(\eta)_i), J(\sigma_g(\eta))_i)$$

$$\leq \varepsilon' + \varepsilon'$$

$$= 2\varepsilon'$$

for $i \in \mathbb{Z}$. Therefore,

$$D((\sigma_f \circ J)\eta, (J \circ \sigma_g)\eta) = \sum_{i \in \mathbb{Z}} 2^{-|i|} d((\sigma_f \circ J(\eta))_i, (J \circ \sigma_g(\eta))_i) \leq 6\varepsilon' \leq \Delta'$$

for every $\eta \in Y_g$, whence

$$D(\sigma_f \circ J, J \circ \sigma_g) \leq \Delta'.$$

This completes the proof. $\qquad\square$

3.4   Application to Stability Theory

The currently most used notions of stability in dynamics are the *Lyapunov, structural* and *topological stabilities* [1, 2, 52, 76]. The Lyapunov stability has a semilocal character and is used in the context of equilibrium points of flows, semiflows and discrete-time dynamical systems. On the other hand, Anosov proved that every U-system (nowadays called an Anosov diffeomorphism) on a closed manifold is structurally stable [2]. Walters proved that Anosov diffeomorphisms are topologically stable [75] too. Nitecki extended Walters's result to all Axiom A diffeomorphisms with the strong transversality condition [56]. Subsequently Walters' proved his stability theorem [76], which asserts that any expansive map with the shadowing property of a compact metric space is topologically stable. Several extensions of topological stability have been proposed over true. One is the following notion of stability due to Sun [71]. It appeared in response to the non-stability of certain surjective Anosov maps on closed topological manifolds.

Definition 3.2 A surjective continuous map $f : X \to X$ is *orbit shift topologically stable* if for every $\varepsilon > 0$, there is $\delta > 0$ such that for every surjective continuous map $g : X \to X$ with $d(f, g) < \delta$, there is $H : X_g \to X_f$ continuous satisfying $D(H(\xi), \xi) \le \varepsilon$ for every $\xi \in X_g$ and $\sigma_f \circ H = H \circ \sigma_g$.

By Theorem 2.3, every expansive homeomorphism with the shadowing property of a compact metric space is topologically GH-stable. Now we merge the notion of topological GH-stability with the previous concept to obtain the following definition.

Definition 3.3 A surjective continuous map $f : X \to X$ of a compact metric space is *orbit shift topologically GH-stable* if for every $\varepsilon > 0$, there is $\delta > 0$ such that for every surjective continuous map $g : Y \to Y$ of a compact metric space satisfying $d_{\mathrm{GH}^0}(f, g) < \delta$, there is a continuous ε-isometry $H : Y_g \to X_f$ such that $\sigma_f \circ H = H \circ \sigma_g$.

With this definition, we obtain the following result.

Theorem 3.2 *Every topologically Anosov map of a compact metric space with positive diameter is orbit shift topologically GH-stable.*

Proof Let f be a topologically Anosov map of a compact metric space with positive diameter. It follows from Theorems 2.2.29 and 2.3.8 in [3] that $\sigma_f : X_f \to X_f$ is expansive and has the shadowing property. Then σ_f is topologically GH-stable by Theorem 2.3.

Now take $\varepsilon > 0$ and let $\Delta > 0$ be given by the topological GH-stability of σ_f. For this Δ, we choose δ from Theorem 3.1. Then if $g : Y \to Y$ is a surjective continuous map of a compact metric space and $d_{\mathrm{GH}^0}(f, g) \le \delta$, then $d_{\mathrm{GH}^0}(\sigma_f, \sigma_g) \le \Delta$. Thus, by the topological GH-stability of f, there is a continuous ε-isometry $H : Y_g \to X_f$ such that $\sigma_f \circ H = H \circ \sigma_g$. This completes the proof.   □

As applications, we will present two examples. The first one is based on the following definition. Given a continuous map $f : X \to X$, $\varepsilon > 0$, and $\xi \in X_f$, we define

$$W_\varepsilon^s(\xi) = \{z_0 \in X : d(f^n(\xi_0), f^n(z_0)) = d(\xi_n, f^n(z_0)) \leq \varepsilon \text{ for all } n \geq 0\}$$

and

$$W_\varepsilon^u(\xi) = \{z_0 \in X : \text{there exists } \eta \in X_f \text{ such that } \pi(\eta) = z_0$$

$$\text{and } d(x_{-n}, \eta_{-n}) \leq \varepsilon \text{ for all } n \geq 0\}.$$

Definition 3.4 A continuous map $f : X \to X$ is *Anosov [69]* if there is $c > 0$ such that for every $0 < \varepsilon < c$, there is $\delta > 0$ such that the intersection $W_\varepsilon^s(\xi) \cap W_\varepsilon^u(\eta)$ consists of exactly one point for every $\xi, \eta \in X_f$ with $d(\xi_0, \eta_0) \leq \delta$.

Sun [71] proved that every surjective Anosov map of a compact metric space is orbit shift topologically stable. This motivates the following result.

Theorem 3.3 *Every surjective Anosov map of a compact metric space is orbit shift topologically GH-stable.*

Proof Let $f : X \to X$ be a surjective continuous map of a compact metric space. If f is Anosov, then f is surjective and c-expansive with the shadowing property by Lemma 1 in [72, p. 170]. Therefore, f is topologically Anosov and thus is orbit shift topologically GH-stable by Theorem 3.2. □

For the next result, we need the following lemma.

Lemma 3.4 *If $h : Y \to X$ is a continuous map of compact metric spaces, then the map $H : Y^{\mathbb{Z}} \to X^{\mathbb{Z}}$ defined by $H(\xi)_i = h(\xi_i)$ for $i \in \mathbb{Z}$ and $\xi \in Y^{\mathbb{Z}}$ is continuous.*

Proof Fix $\rho > 0$ and take $N \geq 0$ such that

$$\sum_{i=N}^{\infty} 2^{-i} \leq \frac{\rho}{4 \operatorname{diam}(X)}.$$

Fix $\gamma > 0$ such that if $\xi_{-N+1}, \xi_{-N+2}, \xi_1, \ldots, \xi_{N-1}, \eta_{-N+1}, \eta_{-N+2}, y_1, \ldots, \eta_{N-1} \in Y$ and

$$\sum_{i=-N+1}^{N-1} 2^{-|i|} d(\xi_i, \eta_i) \leq \gamma,$$

then

$$\sum_{i=-N+1}^{N-1} 2^{-|i|} d(h(\xi_i), h(\eta_i)) \leq \frac{\rho}{2}.$$

Now, suppose that $\xi, \eta \in Y^{\mathbb{Z}}$ satisfy $D(\xi, \eta) \leq \gamma$. Then

$$\sum_{i=-N+1}^{N-1} 2^{-|i|} d(\xi_i, \eta_i) \leq D(\xi, \eta) \leq \rho,$$

and so,

$$\sum_{i=-N+1}^{N-1} 2^{-|i|} d(h(\xi_i), h(\eta_i)) \leq \frac{\rho}{2}.$$

It follows that

$$D(H(\xi), H(\eta)) = \sum_{i=-N+1}^{N-1} 2^{-|i|} d(h(\xi_i), h(\eta_i)) + \sum_{i=N}^{\infty} 2^{-i} d(h(\xi_i), h(\eta_i))$$

$$+ \sum_{i=-\infty}^{N} 2^{i} d(h(\xi_i), h(\eta_i))$$

$$\leq \frac{\rho}{2} + 2 \operatorname{diam}(X) \sum_{i=N}^{\infty} 2^{-(i+1)} = \rho.$$

Hence, H is continuous. □

We can now prove.

Theorem 3.4 *Let $f : X \to X$ be a surjective continuous map of a compact metric space. If f is topologically GH-stable, then f is orbit shift topologically GH-stable.*

Proof Recall that by the diameter of a subset $A \subset X$ is $\operatorname{diam}(A) = \sup\{d(a, a') : a, a' \in A\}$. Since X has more than one point, $\operatorname{diam}(X) > 0$. Fix $\varepsilon > 0$ and take $N \in \mathbb{N}$ such that

$$\sum_{|i| \geq N+1} 2^{-|i|} \leq \frac{\varepsilon}{3 \operatorname{diam}(X)}.$$

For this N, we choose $0 < \beta < \frac{\varepsilon}{3}$ such that

$$d(a, b) \leq \beta \text{ for } a, b \in X \quad \text{implies} \quad d(f^k(a), f^k(b)) \leq \frac{2\varepsilon}{9} \text{ for } 0 \leq k \leq 2N.$$

For this β, we fix $\delta > 0$ from the topological GH-stability of f.

Now take a surjective continuous map $g : Y \to Y$ of a compact metric space such that $d_{\mathrm{GH}^0}(f, g) \leq \delta$. Then there is a continuous β-isometry $h : Y \to X$ such that $f \circ h = h \circ g$. Take $H : Y^{\mathbb{Z}} \to X^{\mathbb{Z}}$ as in Lemma 3.4 for this h. Then H is continuous. If $\xi \in Y_g$, then $f(h(\xi_i)) = h(g(\xi_i)) = h(\xi_{i+1})$ for every $i \in \mathbb{Z}$. So, $H(\xi) \in X_f$, proving $H(Y_g) \subset X_f$. Therefore, we obtain a continuous map $H : Y_g \to X_f$. Since

$$\sigma_f(H(\xi))_i = H(\xi)_{i+1} = h(\xi_{i+1}) = h(\sigma_g(\xi)_i) = H(\sigma_g(\xi))_i$$

for $i \in \mathbb{Z}$ and $\xi \in Y_g$, one has that $\sigma(H(\xi)) = H(\sigma_g(\xi))$ for $\xi \in Y_g$, proving $\sigma_f \circ H = H \circ \sigma_g$.

It remains to verify that H is an ε-isometry. If $\xi, \xi' \in Y_g$, then

$$D(H(\xi), H(\xi')) = \sum_{i \in \mathbb{Z}} 2^{-|i|} d(H(\xi)_i, H(\xi')_i)$$

$$= \sum_{i \in \mathbb{Z}} 2^{-|i|} d(h(\xi_i), h(\xi'_i))$$

$$\leq 3\beta + D(\xi, \xi')$$

$$\leq \varepsilon + D(\xi, \xi'),$$

proving that $D(H(\xi), H(\xi')) - D(\xi, \xi') \leq \varepsilon$. Interchanging the roles of $H(\xi)$ (resp., $H(\xi')$) and ξ (resp., ξ') in this argument, we get $D(\xi, \xi') - D(H(\xi), H(\xi')) \leq \varepsilon$ (see the proof of Lemma 3.1). Therefore,

$$\sup_{\xi, \xi' \in Y_g} |D(H(\xi), H(\xi')) - D(\xi, \xi')| \leq \varepsilon. \tag{3.4}$$

Now take $\eta \in X_f$. Since h is a β-isometry, there is $y \in Y$ such that $d(h(y), \eta_{-N}) \leq \beta$. Define $\xi_{-N+k} = g^k(y)$ for $k \geq 0$. This yields a sequence $\{\xi_i\}_{i \geq -N}$. Since g is surjective, we can extend this sequence for $i < -N$ to get $\xi \in Y_g$. It follows that

$$d(h(\xi_i), \eta_i) = d(f^{i+N}(h(y)), f^{i+N}(\eta_{-N})) \leq \frac{2\varepsilon}{9}$$

for any $-N \leq i \leq N$. Then

$$D(H(\xi), \eta) = \sum_{i \in \mathbb{Z}} 2^{-|i|} d(H(\xi)_i, \eta_i)$$

$$\leq diam(X) \left(\sum_{|i| \geq N+1} 2^{-|i|} \right) + \sum_{i=-N}^{N} 2^{-|i|} d(h(\xi_i), \eta_i)$$

$$\leq \frac{\varepsilon}{3} + \frac{2\varepsilon}{9} \cdot 3 = \varepsilon.$$

Therefore, $D_{\mathrm{H}}(H(Y_g), X_f) \leq \varepsilon$, which, together with (3.4), proves that H is an ε-isometry, as needed.   □

Exercises

Exercise 3.5 Prove that the tent map $f(x) = \max\{2x, 1 - 2x\}$ for $x \in [0, 1]$ is neither topologically GH-stable, nor orbit shift topologically GH-stable.

Exercise 3.6 Is it true that if $f : X \to X$ is a *homeomorphism* of a compact metric space, then f is orbit shift topologically GH-stable if and only if it is topologically GH-stable?

Exercise 3.7 Is Theorem 3.1 true when $\mathrm{diam}(X) = 0$, i.e., when X is a one-point set?

Shadowing from the Gromov-Hausdorff Viewpoint 4

4.1   Introduction

When simulating a given system, it is important to know under which conditions approximated trajectories may be tracked by real ones. If this is the case for all approximated trajectories, then we say that the system has the *shadowing property*. This property was discovered by Bowen [62], who proved it for hyperbolic systems. Since then, several generalizations were proposed.

In this section, we introduce one more scale generalization treat takes into account Gromov-Hausdorff perturbations of the underlying systems. This leads two kinds of shadowing, namely, the Gromov-Hausdorff and the weak Gromov-Hausdorff shadowing properties. As the names indicate, every system with the Gromov-Hausdorff shadowing property has the weak Gromov-Hausdorff shadowing property. We show that every topologically GH-stable system has the weak Gromov-Hausdorff shadowing property, that the shadowing property implies the Gromov-Hausdorff shadowing property, and finally that every expansive system with the weak Gromov-Hausdorff shadowing property is topologically GH-stable.

4.2   Definitions, Statement of Main Results, and Proofs

Hereafter, X will denote a compact metric space with more than one point. The definition of shadowing property motivates the following one. We say that a homeomorphism $f : X \to X$ of a compact metric space has the *Gromov-Hausdorff shadowing property* (or *GH- shadowing property* for short) if for every $\varepsilon > 0$, there is $\delta_* > 0$ such that for every $0 < \delta < \delta_*$ and every homeomorphism $g : Y \to Y$ of a compact metric space with

© The Author(s), under exclusive license to Springer Nature Switzerland AG 2022
J. Lee, C. Morales, *Gromov-Hausdorff Stability of Dynamical Systems and Applications to PDEs*, Frontiers in Mathematics, https://doi.org/10.1007/978-3-031-12031-2_4

$d_{GH^0}(f, g) \leq \delta$, there is a δ-isometry $j : Y \rightarrow X$ such that for every δ-pseudo orbit $\{y_n\}_{n \in \mathbb{Z}}$ of g, there is $x \in X$ such that $d(f^n(x), j(y_n)) \leq \varepsilon$ for every $n \in \mathbb{Z}$.

In the recent paper [9], the following variation of shadowing was proposed: We say that f has the *weak shadowing property* if for every $\varepsilon > 0$, there is $\delta > 0$ such that for every homeomorphism $g : X \rightarrow X$ with $d(f, g) \leq \delta$ and every $y \in X$, there is $x \in X$ such that $d(f^n(x), g^n(y)) \leq \varepsilon$ for every $n \in \mathbb{Z}$. This definition motivates the following concept.

Definition 4.1 We say that a homeomorphism $f : X \rightarrow X$ of a compact metric space has the *weak Gromov-Hausdorff shadowing property* (or the *weak GH-shadowing property* for short) if for every $\varepsilon > 0$, there is $\delta_* > 0$ such that for every $0 < \delta < \delta_*$ and every homeomorphism $g : Y \rightarrow Y$ of a compact metric space with $d_{GH^0}(f, g) \leq \delta$, there is a δ-isometry $j : Y \rightarrow X$ such that for every $y \in Y$, there is $x \in X$ such that $d(f^n(x), j(g^n(y))) \leq \varepsilon$ for every $n \in \mathbb{Z}$.

The difference between GH-shadowing and weak GH-shadowing is that the former deals with general δ-pseudo orbits $(y_n)_{n \in \mathbb{Z}}$, while the latter deals with true orbits $(g^n(y))_{n \in \mathbb{Z}}$. In particular, every homeomorphism with the GH-shadowing property has the weak GH-shadowing property. We shall prove the following facts for every homeomorphism $f : X \rightarrow X$ of a compact metric space.

Proposition 4.1 *If f has the shadowing property, then f has the GH-shadowing property.*

Proof Let $\varepsilon > 0$ and $\delta' > 0$ be given by the shadowing property of f for this ε. Set $\delta_* = \frac{\delta'}{3}$ and take $0 < \delta < \delta_*$. If $g : Y \rightarrow Y$ is a homeomorphism of a compact metric space such that $d_{GH^0}(f, g) \leq \delta$, then there is a δ-isometry $j : Y \rightarrow X$ such that $d(f \circ j, j \circ g) \leq \delta$. Let $(y_n)_{n \in \mathbb{Z}}$ be any δ-pseudo orbit of g. Since for every $n \in \mathbb{Z}$,

$$d(f(j(y_n)), j(y_{n+1})) \leq d(f \circ j(y_n), j \circ g(y_n)) + d(j(g(y_n)), j(y_{n+1}))$$

$$\leq d(f \circ j, j \circ g) + \delta + d(g(y_n), y_{n+1})$$

$$\leq 3\delta < 3\delta_* = \delta',$$

$(j(y_n))_{n \in \mathbb{Z}}$ is a δ'-pseudo orbit of f. Then there is $x \in X$ such that $d(f^n(x), j(y_n)) \leq \varepsilon$ for every $n \in \mathbb{Z}$. This ends the proof.   □

It is natural to ask if the converse of Proposition 4.1 holds.

Question 4.1 Does the GH-shadowing property imply the shadowing property?

A partial answer we provide uses the following auxiliary shadowing concept:

Definition 4.2 We say that f has the *isometric shadowing property* if for every $\varepsilon > 0$, there is $\delta_* > 0$ such that for every $0 < \delta < \delta_*$, there is a δ-isometry $j : X \to X$ such that for every δ-pseudo orbit $(x_n)_{n \in \mathbb{Z}}$ of f, there is $x \in X$ such that $d(f^n(x), j(x_n)) \leq \varepsilon$ for every $n \in \mathbb{Z}$.

Proposition 4.2 *If f has the GH-shadowing property, then f has the isometric shadowing property.*

Proof Given $\varepsilon > 0$, we take δ_* as provided by the GH-shadowing property. If $0 < \delta < \delta_*$, then we can take $g = f$ to get $d_{\mathrm{GH}^0}(f, g) = 0 < \delta$. So, there is a δ-isometry $j : X \to X$, as desired. $\square$

The following fact is easy to prove.

Proposition 4.3 *If f has the shadowing property, then f has the isometric shadowing property.*

Recall that a homeomorphism $f : X \to X$ is *expansive* if there is $e > 0$ (called an expansivity constant) such that if $x, y \in X$ and $d(f^n(x), f^n(y)) \leq e$ for every $n \in \mathbb{Z}$, then $x = y$.

Proposition 4.4 *If f is expansive and has the weak GH-shadowing property, then f is topologically GH-stable.*

Proof Let e be an expansivity constant of f. Take $\varepsilon > 0$ and let $0 < \varepsilon' < \frac{1}{8} \min\{e, \varepsilon\}$. For this ε', we let δ_* be given by the weak GH-shadowing property of f. Take $0 < \delta < \frac{1}{3} \min\{\delta_*, \varepsilon'\}$ and a homeomorphism $g : Y \to Y$ of a compact metric space such that $d_{\mathrm{GH}^0}(f, g) \leq \delta$.

Let $j : Y \to X$ be the δ-isometry given by the weak GH-shadowing property for this g. Then for $y \in Y$, there is $x \in X$ such that

$$d(f^n(x), j(g^n(y))) \leq \varepsilon'$$

for all $n \in \mathbb{Z}$. This x is unique: if there is an x' different from x, then

$$d(f^n(x), f^n(x')) \leq d(f^n(x), j(g^n(y))) + d(f^n(x'), j(g^n(y))) \leq 2\varepsilon' < e,$$

and so $x = x'$ by expansivity.

From this uniqueness, we get a map $h : Y \to X$ such that

$$d(f^n(h(y)), j(g^n(y))) \leq \varepsilon' \tag{4.1}$$

for $y \in Y$ and $n \in \mathbb{Z}$. Taking here $n = 1$, we get

$$d(h(y), j(y)) \leq \varepsilon'.$$

Therefore,

$$d(h(y), h(y')) \leq d(h(y), j(y)) + \delta + d(y, y') + d(j(y'), h(y')) < \varepsilon + d(y, y')$$

and also,

$$d(y, y') < \delta + d(j(y), h(y)) + d(h(y), h(y')) + d(j(y'), h(y')) < \varepsilon + d(h(y), h(y'))$$

for all $y, y' \in Y$. This implies that

$$\sup_{y, y' \in Y} |d(h(y), h(y')) - d(y, y')| < \varepsilon.$$

Moreover, for $x \in X$ there is $y \in Y$ such that $d(j(y), x) < \delta$. So,

$$d(h(y), x) \leq d(h(y), j(y)) + d(j(y), x) \leq \varepsilon' + \delta < 2\varepsilon' < \varepsilon.$$

Hence, $d_H(h(Y), X) < \varepsilon$, proving that h is an ε-isometry.

By replacing n by $n + 1$ and y by $g(y)$ in (4.1), we obtain

$$d(f^n(f(h(y))), j(g^{n+1}(y)) \leq \varepsilon'$$

and

$$d(f^n(g(h(y))), j(g^{n+1}(y))) \leq \varepsilon'$$

for all $n \in \mathbb{Z}$, respectively. It follows that

$$d(f^n(f(h(y))), f^n(h(g(y)))) \leq 2\varepsilon' < e$$

for all $n \in \mathbb{Z}$. Since e is an expansivity constant of f, we get $f(h(y)) = g(h(y))$ for every $y \in Y$, proving $f \circ h = h \circ g$.

Finally, we show that h is continuous. Fix $\Delta > 0$. Since e is an expansivity constant of f, there is $N \in \mathbb{N}$ such that

$$a, b \in X \text{ and } d(f^n(a), f^n(b)) \leq e \text{ for } -N \leq n \leq N \quad \text{implies} \quad d(a, b) \leq \Delta$$

(cf. [76]). Since g is continuous and Y compact, there is $\gamma > 0$ such that

$$d(y, y') \leq \gamma \text{ for } y, y' \in Y \quad \text{implies} \quad d(g^n(y), g^n(y')) \leq \frac{e}{3} \text{ for } -N \leq n \leq N.$$

Then if $y, y' \in Y$ and $d(y, y') \leq \gamma$, we have

$$d(f^n(h(y)), f^n(h(y'))) = d(h(g^n(y)), h(g^n(y')))$$
$$\leq d(h(g^n(y)), j(g^n(y))) + d(j(g^n(y)), j(g^n(y'))) +$$
$$d(h(g^n(y')), j(g^n(y')))$$
$$< 2\varepsilon' + \delta + d(g^n(y), g^n(y'))$$
$$< 3\varepsilon' + \frac{e}{3} < \frac{10e}{24} < e$$

for $-N \leq n \leq N$, and so $d(h(y), h(y')) \leq \Delta$. Then h is continuous, which completed the proof follows.   □

Next we will prove the following fact. Recall that a metric space X is *totally disconnected* if every compact connected subset of X consists of a single point.

Proposition 4.5 *The identity* Id $: X \to X$ *has the isometric shadowing property if and only if X is totally disconnected.*

Proof If X is totally disconnected, then Id has the shadowing property (Theorem 2.3.2 in [3]). So, Id has the isometric shadowing property by Proposition 4.3.

Conversely, suppose that Id has the isometric shadowing property and, by contradiction, that X is not totally disconnected. Then there is a closed connected set F with diameter $\text{diam}(F) > 0$. Take δ_* from the isometric shadowing property of f for $\varepsilon = \frac{\text{diam}(F)}{4}$ and $0 < \delta < \frac{1}{4}\min\{\delta_*, \varepsilon\}$. For this δ, take the δ-isometry $j : X \to X$ from the isometric shadowing property of Id.

Since F is compact, there are points $x, y \in F$ such that $d(x, y) = \text{diam}(F)$. Since F is connected, there exists a sequence $x = x_0, x_1, \ldots, x_N = y$ such that $d(x_n, x_{n+1}) < \delta$ for every $0 \leq n \leq N - 1$. By setting $x_n = x$ for $n \leq 0$ and $x_n = y$ for $n \geq N$, we get a δ-pseudo orbit $(x_n)_{n \in \mathbb{Z}}$ of Id. It follows that there is $x \in X$ such that $d(x, j(x_n)) \leq \varepsilon$ for every $n \in \mathbb{Z}$. Then,

$$\text{diam}(F) = d(x, y)$$
$$< \delta + d(j(x_0), j(x_N))$$
$$\leq \delta + d(x, j(x_0)) + d(x, j(x_N))$$

$$\leq \frac{\mathrm{diam}(F)}{4} + 2\varepsilon$$

$$\leq \frac{3\,\mathrm{diam}(F)}{4} < \mathrm{diam}(F),$$

a contradiction, which completes the proof.   □

From this fact, we get the following.

Proposition 4.6 *The identity* $\mathrm{Id} : X \to X$ *has the GH-shadowing property if and only if* X *is totally disconnected.*

Recall that a homeomorphism $f : X \to X$ is *minimal* if the orbit $O_f(x) = (f^n(x))_{n \in \mathbb{Z}}$ is dense in X for every $x \in X$.

Proposition 4.7 *If* $f : X \to X$ *is a minimal homeomorphism of a compact connected metric space with more than one point, then* f *has neither the GH-shadowing property, nor the isometric shadowing property.*

Proof It suffices to show that f does not have the isometric shadowing property. Suppose, by contradiction, that it does. Since X is not a point,

$$\mathrm{diam}(X) > 0.$$

Let δ_* be given by the isometric shadowing property of f for $\varepsilon = \frac{\mathrm{diam}(X)}{4}$. Take $0 < \delta < \frac{1}{4}\min\{\delta_*, \varepsilon\}$ and the corresponding δ-isometry $j : X \to X$.

Take any $x_0 \in X$. Since f is minimal, there is $k \in \mathbb{N}$ such that $f^k(x_0) \in B(x_0, \delta)$. Defining

$$x_{mk+i} = f^i(x_0)$$

for $m \in \mathbb{Z}$ and $0 \leq i < k$, we obtain a δ-pseudo orbit $(x_n)_{n \in \mathbb{Z}}$ of f. Then, by isometric shadowing, there is $x \in X$ such that

$$d(f^n(x), j(x_n)) \leq \varepsilon$$

for all $n \in \mathbb{Z}$. Taking here $n = mk$ for $m \in \mathbb{Z}$, we obtain $d((f^k)^m(x), j(x_0)) \leq \varepsilon$ for every $m \in \mathbb{Z}$. So,

$$O_{f^k}(x) \subset B[x_0, \varepsilon].$$

But f is minimal and X is connected, so f^k is minimal. Then we have $X = \overline{O_{f^k}(x)}$ and $X \subset B[x_0, \varepsilon]$. Thus,

$$\mathrm{diam}(X) \leq 2\varepsilon = \frac{\mathrm{diam}(X)}{2}$$

a contradiction, the proof in complete.   □

The following property can be proved as in the shadowing property case [3].

Proposition 4.8 f has the GH-*shadowing (resp., isometric shadowing) property if and only if* f^k *has such a property for some* $k \in \mathbb{N}$.

We use it in the following example.

Example 4.2 Proposition 4.7 implies that irrational rotations of the circle have neither the GH-shadowing property, nor the isometric shadowing property. Since $f^k = I$ for some $k \in \mathbb{N}$ when f is a rational rotation of the circle. Propositions 1.4 and 4.8 imply that rational rotations of the circle do not have these properties.

Exercises

The following exercise is motivated by the notion of persistent homeomorphism introduced by Lewowicz [50].

Exercise 4.3 Find an example of a *GH-persistent* homeomorphism, that is, a homeomorphism $f : X \to X$ of a compact metric space such that for every $\varepsilon > 0$, there is $\delta_* > 0$ such that for every homeomorphism $g : X \to X$ of a compact metric space with $d_{\mathrm{GH}^0}(f, g) \leq \delta$ for some $0 < \delta < \delta_*$, there is a δ-isometry $j : Y \to X$ such that for every $x \in X$, there is $y \in Y$ such that $d(f^n(x), j(g^n(y))) \leq \varepsilon$ for every $n \in \mathbb{Z}$.

Exercise 4.4 *(True or false)* An equicontinuous *(Exercise 1.30)* homeomorphism of a compact metric space X has the *GH*-shadowing property if and only if X is totally disconnected.

Part II
Applications to PDEs

Introduction

In the theory as well as in the applications of differential equations in nonlinear science, and especially in engineering and mathematical physics, finding exact solutions used to be one of the major concerns. Exact solutions can be used to analyze the qualitative properties of the equations encountered in such applications. However, in general, finding exact solutions of equations that describes complicated natural phenomena is difficult and often impossible. Even when exact solutions can be found, extracting from them qualitative information on solution curves, such as their asymptotic behavior or the existence of invariant manifolds, etc, is not an easy task.

In the geometric or qualitative theory of differential equations, the main goal is to describe the geometry of solution curves. At the end of the nineteenth century, Poincaré introduced a new method to understand the long-time behavior of solution curves of ordinary differential equations without the need to find exact solutions of the equations under study. Precisely, he used one map to express the collection of all solutions of an equation, map called the dynamical system or the flow generated by the given equation, and investigated the existence of special solutions such as equilibrium points, periodic solutions, recurrent solutions. Since the early 1980s and to the 2000s, many authors, among them Chepyzhov, Glendinning, Hale, Henry, Sell, Robinson, Temam, Vishik, Bo You, initiated and deepened the study of the qualitative properties of solutions of partial differential equations, in a setting that today is called the geometric theory of partial differential equations [20, 34, 64, 65, 73]. In this theory questions concerning stability play an important role.

There are several ways to discuss and characterize the stability of solutions of differential equations. The most useful tools in this respect employ objects called global attractors and inertial manifolds. A global attractor is a compact invariant set which attracts all solution curves; an inertial manifold is an invariant manifold (non-compact, in general) which attracts all solution curves exponentially. Often a given partial differential equation can be transformed into an ordinary differential equation in a Banach space. The importance of global attractors and inertial manifolds is that the study of the asymptotic behaviors of solutions of an ordinary differential equation in an infinite-dimensional space can be reduced to a study in a finite-dimensional setting.

Here we are interested in studying the behavior of global attractor and inertial manifold of reaction-diffusion equations with respect to perturbations of the domain and of the equation (compare with perturbation theory of linear operators [41]). Recently a lot of interesting results in this direction have been obtained by many researchers (see for example [4, 6, 11, 51]).

Let us consider a reaction-diffusion equation of the form

$$\begin{cases} \partial_t u - \Delta u = f(u) & \text{in } \Omega \times (0, \infty), \\ u = 0 & \text{on } \partial\Omega \times (0, \infty). \end{cases} \tag{1}$$

For the problem of domain perturbation, one of the difficulties is that the phase space of the induced dynamical system changes as we change the domain. In fact, the phase spaces $H_0^1(\Omega)$ and $H_0^1(\Omega_\varepsilon)$ which contain global attractors $\mathcal{A}$ and $\mathcal{A}_\varepsilon$, respectively, can be disjoint even if Ω_ε is a small perturbation of $\Omega \subset \mathbb{R}^n$.

To overcome this difficulty, many people used the technique adopted their by Henry in [34, 36] which makes it possible to consider the problem of continuity of the attractors in the fixed phase space $H^1(\Omega_0)$ as $\Omega_\varepsilon \to \Omega_0$ (e.g., see [11, 60]). More precisely, they followed the general approach, which basically consists in "pull-backing" the perturbed problems to the fixed domain Ω_0 and then considering the family of abstract semilinear problems thus generated. Arrieta and Carvalho [6] applied another method, which uses extended phase spaces to compare attractors living in different phase spaces. More precisely, they considered the extended phase space

$$H_\varepsilon^1 := H^1(\Omega \cap \Omega_\varepsilon) \oplus H^1(\Omega \setminus \Omega_\varepsilon) \oplus H^1(\Omega_\varepsilon \setminus \Omega)$$

with the norm $\| \cdot \|_{H_\varepsilon^1}$. However, the norms $\| \cdot \|_{H_\varepsilon^1}$ and $\| \cdot \|_{H_\delta^1}$ on the extended spaces H_ε^1 and H_δ^1, respectively, are in general not comparable.

In this part, we will use the Gromov-Hausdorff distance between two global attractors and between two inertial manifolds which belong to disjoint phase spaces to derive their continuity and the Gromov-Hausdorff stability under perturbations of the domain and equation. The notion of Gromov-Hausdorff distance defined on the collection of metric spaces was first introduced by Edwards [27] and was developed by Fukaya et al. [16, 30] to study the convergence and collapsing of Riemannian manifolds. Arbieto and Morales [5] investigated the stability of expansive maps with the shadowing property under the Gromov-Hausdorff topology.

We note that pull-backing or domain extension methods do not consider an appropriate topology to compare two global attractors in different phase spaces. Moreover, these methods do not provide information on the change of the dynamics of solution curves inside the global attractor under the perturbations being considered. The Gromov-Hausdorff distance induces a topology on the collection of all metric spaces up to isometry, and as such it gives us a strong tool to compare our objects in different phase spaces. Moreover, this distance applied to dynamical systems allows us to get information on the internal dynamics under perturbations of the domain.

To study the problem of perturbation of the equation, many researchers transform Eq. (1) into an ordinary differential equation on a Banach space, e.g., Lebesgue space L^p, Sobolev space like $W^{1,p}$ or fractional Sobolev space. For example, consider the transformed equation on Banach space $L^2(\Omega)$

$$u_t + Au = F(u) \text{ for } u \in L^2(\Omega) \tag{2}$$

where A is $-\Delta$ with Dirichlet boundary condition, and F is the Nemytskii operator of f. Many researchers assumed that Eq. (2) induces the C^1 dynamical systems and applied the methods which have been developed in the theory of differential dynamical systems such as hyperbolicity, transversality and Morse-Smale systems, to study the internal structure of solutions of Eq. (2). However when dealing with evolution equations in L^p spaces, there are no differentiable functions of Nemytskii type from L^p into itself other than affine functions. We note that the theory of stability so far only allows us to consider differentiable dynamical systems with C^1 perturbations which is far from being a reality in L^p spaces.

The content of this part is as follows. In Chap. 5, we study the residual continuity of the global attractors with respect to the Gromov-Hausdorff topology and their dynamics for reaction-diffusion equations on finite-dimensional domains. In Chap. 6, we analyze the Gromov-Hausdorff stability and continuous dependence of the inertial manifolds under perturbations of the domain and of the equation. Finally, in Chap. 7, we investigate the geometric theory of the Chafee-Infante equations under Lipschitz perturbations.

GH-Stability of Reaction-Diffusion Equations

5

5.1   Introduction

In this chapter, we use the Gromov-Hausdorff distances between two global attractors (that belong to disjoint phase spaces) and two dynamical systems to consider the continuous dependence of the global attractors and the stability of the dynamical systems on global attractors induced by a reaction-diffusion equation under perturbations of the domain. The novelty of the method is that one compares any two systems in different phase spaces without the need to "pull-back" the perturbed systems to the original domain.

Let Ω be an open bounded domain in $\mathbb{R}^N$ ($N \geq 2$) with smooth boundary. Consider the following (initial-boundary value problem for a) reaction-diffusion equation:

$$\begin{cases} u_t - \Delta u + f(u) = g & \text{in } \Omega \times (0, \infty), \\ u = 0 & \text{on } \partial\Omega \times (0, \infty), \\ u(x, 0) = u_0 & \text{in } \Omega. \end{cases} \tag{5.1}$$

Assume that $u_0 \in L^2(\Omega)$, $g \in L^2(\mathbb{R}^N) \cap L^\infty(\mathbb{R}^N)$, and $f : \mathbb{R} \to \mathbb{R}$ is a C^1 function satisfying the following condition: there exist four positive constants c_0, c_1, c_2, ℓ such that

$$c_1|s|^p - c_0 \leq f(s)s \leq c_2|s|^p + c_0 \quad \text{with } p \geq 2 \text{ for all } s \in \mathbb{R}, \tag{5.2}$$

$$f'(s) \geq -\ell. \tag{5.3}$$

It is well known that the problem (5.1) is well posed in various function spaces and has a global attractor in $L^2(\Omega)$ (see, for example, [54, 80]).

© The Author(s), under exclusive license to Springer Nature Switzerland AG 2022
J. Lee, C. Morales, *Gromov-Hausdorff Stability of Dynamical Systems and Applications to PDEs*, Frontiers in Mathematics, https://doi.org/10.1007/978-3-031-12031-2_5

We let Diff(Ω) denote the space of diffeomorphisms h from Ω onto its image $\Omega_h :=$ $h(\Omega) \subset \mathbb{R}^N$ with the C^1 topology.

For any $h \in \text{Diff}(\Omega)$, let us consider the following reaction-diffusion equation:

$$
\begin{cases}
u_t - \Delta_h u + f(u) = g & \text{in } \Omega_h \times (0, \infty), \\
u = 0 & \text{on } \partial\Omega_h \times (0, \infty), \\
u(x, 0) = u_0 & \text{in } \Omega_h.
\end{cases}
\tag{5.4}
$$

Here, Δ_h denotes the Laplacian operator on Ω_h with homogeneous Dirichlet boundary condition.

Let $S_h : L^2(\Omega_h) \times \mathbb{R}_0^+ \to L^2(\Omega_h)$ be the semidynamical system induced by (5.4), where $\mathbb{R}_0^+ = [0, \infty)$. Denote by $\mathcal{A}_h$ the global attractor of S_h. Note that S_h is injective on $\mathcal{A}_h$ when $g \in L^\infty(\Omega_h)$. Hence, we can see that the restriction of the semidynamical system S_h to $\mathcal{A}_h$ gives rise to a dynamical system, which will be denoted by $\phi_h : \mathcal{A}_h \times \mathbb{R} \to \mathcal{A}_h$.

A subset $\mathcal{R}$ of a topological space X is said to be *residual* if $\mathcal{R}$ contains a countable intersection of open and dense subsets of X. We say that a map $f : X \to Y$ between topological spaces is *residually continuous* if there exists a residual subset $\mathcal{R}$ of X such that f is continuous at every point of $\mathcal{R}$.

In this section, we prove the following two theorems.

Theorem 5.1 *The map* $\mathbb{A} : \text{Diff}(\Omega) \to \mathcal{M}$ *given by* $\mathbb{A}(h) = \mathcal{A}_h$ *is residually continuous, where* $\mathcal{A}_h$ *is the global attractor of the semidynamical system* S_h *induced by* (5.4).

Theorem 5.2 *The global attractor* $\mathcal{A}_h$ *of the semidynamical system* S_h *induced by* (5.4) *is residually GH-stable with respect to* $h \in \text{Diff}(\Omega)$. *More precisely, the map* $\phi : \text{Diff}(\Omega) \to$ *CDS given by* $\phi(h) = \phi_h$, *where* ϕ_h *is the dynamical system on* $\mathcal{A}_h$, *is residually continuous.*

5.2   Proof of Theorem 5.1

For any $h \in \text{Diff}(\Omega)$, we let $\Omega_h := h(\Omega)$ and Δ_h be the Laplacian operator on Ω_h. If we denote by $\lambda_1(\Omega_h)$ the first eigenvalue of Δ_h with the domain $H^2(\Omega_h) \cap H_0^1(\Omega_h)$, then the Poincaré inequality

$$
\|\nabla_h u\|_{L^2(\Omega_h)}^2 \geq \lambda_1(\Omega_h)\|u\|_{L^2(\Omega_h)}^2
\tag{5.5}
$$

holds for all $u \in H_0^1(\Omega_h)$, where ∇_h denotes the gradient on Ω_h. Note that $\lambda_1(\Omega_h) \geq \lambda_1(\Omega_{\tilde{h}})$ if Ω_h is a subset of $\Omega_{\tilde{h}}$, for $h, \tilde{h} \in \text{Diff}(\Omega)$. In view of the inequality (5.5), we can consider the space $H_0^1(\Omega_h)$ with the norm $\|u\|_{H_0^1(\Omega_h)} := \|\nabla_h u\|_{L^2(\Omega_h)}$.

We first recall the well-posedness of the problem (5.4) in the such weak solutions, which will be used in the proof of Theorem 5.1.

Lemma 5.1 ([54, 80]) *The problem* (5.4) *has a unique weak solution u which depends continuously on the initial datum $u_0 \in L^2(\Omega_h)$ such that for any $T > 0$,*

$$u \in C([0, T]; L^2(\Omega_h)) \cap L^2(0, T; H^1_0(\Omega_h)) \cap L^p(0, T; L^p(\Omega_h)),$$

$$\frac{\partial u}{\partial t} \in L^2(0, T; H^{-1}(\Omega_h)) + L^q(0, T; L^q(\Omega_h)),$$

where $q = p/(p-1)$ is the conjugate index of $p \geq 2$. Moreover, every weak solution u of (5.4) *satisfies*

$$\|u(t)\|^2_{L^2(\Omega_h)} \leq \frac{1}{\lambda_1(\Omega_h)} \left(2c_0|\Omega_h| + \frac{\|g\|^2_{L^2(\Omega_h)}}{\lambda_1(\Omega_h)} \right) \cdot (1 - e^{-2\lambda_1(\Omega_h)t}) \tag{5.6}$$

$$+ e^{-2\lambda_1(\Omega_h)t} \|u_0\|^2_{L^2(\Omega_h)} \quad \text{for all } t \geq 0,$$

$$\|u(t)\|^2_{L^2(0,T;H^1_0(\Omega_h))} + 2c_1 \|u\|^p_{L^p(0,T;L^p(\Omega_h))} \tag{5.7}$$

$$\leq 2c_0|\Omega_h|T + \frac{T\|g\|^2_{L^2(\Omega_h)}}{\lambda_1(\Omega_h)} + \|u_0\|^2_{L^2(\Omega_h)}, \quad \text{for all } t \geq 0$$

and

$$\|f(u)\|^q_{L^q(0,T;L^q(\Omega_h))} \leq \tilde{c}_2 \left(|\Omega_h|T + \|u\|^p_{L^p(0,T;L^p(\Omega_h))} \right). \tag{5.8}$$

Furthermore, we have a semidynamical system $S_h : L^2(\Omega_h) \times \mathbb{R}^+_0 \to L^2(\Omega_h)$ defined by $S_h(u_0, t) := u(t)$, where $u(t)$ is the unique weak solution of (5.4) *with initial datum u_0, and S_h has a compact absorbing set B_h in $L^2(\Omega_h)$:*

$$B_h := \{u \in H^1_0(\Omega_h) : \|u\|_{H^1_0(\Omega_h)} \leq r_h\} \subset B(0, R_h), \tag{5.9}$$

where r_h and R_h are sufficiently large constants that depend on Ω_h and the constants in (5.2) *and* (5.3).

For any $h_0, h_n \in \text{Diff}(\Omega)$ with $n \in \mathbb{N}$, we consider the map $i_n : L^2(\Omega_{h_n}) \to L^2(\Omega_{h_0})$ defined by

$$i_n(u) = u \circ (h_n \circ h_0^{-1}) \text{ for } u \in L^2(\Omega_{h_n}),$$

and its inverse $i_n^{-1} : L^2(\Omega_{h_0}) \to L^2(\Omega_{h_n})$, given by

$$i_n^{-1}(v) = v \circ (h_0 \circ h_n^{-1}) \text{ for } v \in L^2(\Omega_{h_0}).$$

Then we see that i_n is a bounded isomorphism and the restricted map $i_n : H_0^1(\Omega_{h_n}) \to H_0^1(\Omega_{h_0})$ is also a bounded isomorphism.

We need the following lemma, which will allow us to apply Proposition 2.1 in the proof of Theorem 5.1.

Lemma 5.2 *Let $\{h_n\}_{n\in\mathbb{N}}$ be a sequence in $\mathrm{Diff}(\Omega)$ and $\{v_{n,0}\}_{n\in\mathbb{N}}$ be a sequence in $L^2(\Omega_{h_0})$ such that*

$$h_n \to h_0 \text{ and } v_{n,0} \to v_0 \text{ as } n \to \infty.$$

Then for any $T > 0$, we have

$$\|i_n(S_{h_n}(i_n^{-1}(v_{n,0}), t)) - S_{h_0}(v_0, t)\|_{L^2(\Omega_{h_0})} \to 0 \text{ for all } t \in [0, T].$$

Proof Take $T > 0$ and let $u_{n,0} = i_n^{-1}(v_{n,0})$ for each $n \in \mathbb{N}$. Then $u_n(t) = S_{h_n}(u_{n,0}, t)$ is a weak solution of the problem

$$\begin{cases} \partial_t u_n - \Delta_{h_n} u_n + f(u_n) = g & \text{in } \Omega_{h_n} \times (0, T], \\ u_n = 0 & \text{on } \partial\Omega_{h_n} \times (0, T], \\ u_n(x, 0) = u_{n,0} & \text{in } \Omega_{h_n}, \end{cases} \tag{5.10}$$

and $v(t) = S_{h_0}(v_0, t)$ is a weak solution of the problem

$$\begin{cases} \partial_t v - \Delta_{h_0} v + f(v) = g & \text{in } \Omega_{h_0} \times (0, T], \\ v = 0 & \text{on } \partial\Omega_{h_0} \times (0, T], \\ v(x, 0) = v_0 & \text{in } \Omega_{h_0}. \end{cases} \tag{5.11}$$

Let $v_n = i_n(u_n)$ and $z_n = v_n - v$. We have $u_n \in L^2(0, T; H_0^1(\Omega_{h_n}))$, $v_n \in L^2(0, T; H_0^1(\Omega_{h_0}))$, and $v \in L^2(0, T; H_0^1(\Omega_{h_0}))$. Consequently, $i_n^{-1}(z_n) \in L^2(0, T; H_0^1(\Omega_{h_n}))$. Hence, we can multiply the first equation in (5.10) by $i_n^{-1}(z_n)$, integrate over Ω_{h_n}, and integrate by parts to deduce that

$$\int_{\Omega_{h_n}} (\partial_t u_n)(i_n^{-1}(z_n))dx + \int_{\Omega_{h_n}} \nabla_{h_n} u_n \cdot \nabla_{h_n}(i_n^{-1}(z_n))dx$$

$$= -\int_{\Omega_{h_n}} f(u_n)i_n^{-1}(z_n)dx + \int_{\Omega_{h_n}} g i_n^{-1}(z_n)dx. \tag{5.12}$$

Let $y = h_0 \circ h_n^{-1}(x)$ for $x \in \Omega_{h_n}$. By changing variables and using the notation x instead of y for convenience, we get from (5.12) that

$$\int_{\Omega_{h_0}} (v_n)_t z_n |\det H_n| dx + \int_{\Omega_{h_0}} \overline{H}_n \nabla_{h_0} v_n \cdot \overline{H}_n \nabla_{h_0} z_n |\det H_n| dx$$

$$= -\int_{\Omega_{h_0}} f(v_n) z_n |\det H_n| dx + \int_{\Omega_{h_0}} z_n i_n(g) |\det H_n| dx, \tag{5.13}$$

where

$$H_n := H_n(x) = D(h_n \circ h_0^{-1}(x)) = \left(\frac{\partial (h_n \circ h_0^{-1}(x))_i}{\partial x_j}\right)_{N \times N}$$

for $x \in \Omega_{h_0}$, and $\overline{H}_n = (H_n^{-1})^{\mathsf{T}}$ is the transpose of the inverse H_n^{-1}.

Since $z_n \in L^2(0, T; H_0^1(\Omega_{h_0}))$, we can multiply the first equation in (5.11) by z_n and integrate over Ω_{h_0} to obtain

$$\int_{\Omega_{h_0}} v_t z_n dx + \int_{\Omega_{h_0}} \nabla_{h_0} v \cdot \nabla_{h_0} z_n dx = -\int_{\Omega_{h_0}} f(v) z_n dx + \int_{\Omega_{h_0}} g z_n dx. \tag{5.14}$$

By combining (5.13) and (5.14), we get

$$\int_{\Omega_{h_0}} (z_n)_t z_n |\det H_n| dx + \int_{\Omega_{h_0}} (|\det H_n| - 1) v_t z_n dx + \int_{\Omega_{h_0}} |\overline{H}_n \nabla_{h_0} z_n|^2 |\det H_n| dx$$

$$= \int_{\Omega_{h_0}} (I - \overline{H}_n) \nabla_{h_0} v \cdot \overline{H}_n \nabla_{h_0} z_n |\det H_n| dx \tag{5.15}$$

$$+ \int_{\Omega_{h_0}} \nabla_{h_0} v \cdot (I - \overline{H}_n) \nabla_{h_0} z_n |\det H_n| dx + \int_{\Omega_{h_0}} (1 - |\det H_n|) \nabla_{h_0} v \cdot \nabla_{h_0} z_n dx$$

$$+ \int_{\Omega_{h_0}} (1 - |\det H_n|) f(v) z_n dx - \int_{\Omega_{h_0}} (f(v_n) - f(v)) z_n |\det H_n| dx$$

$$+ \int_{\Omega_{h_0}} (i_n(g) - g) z_n |\det H_n| dx + \int_{\Omega_{h_0}} g z_n (|\det H_n| - 1) dx,$$

where the following identity (and other simple identities) is used:

$$\int_{\Omega_{h_0}} \overline{H}_n \nabla_{h_0} v_n \cdot \overline{H}_n \nabla_{h_0} z_n |\det H_n| dx - \int_{\Omega_{h_0}} \nabla_{h_0} v \cdot \nabla_{h_0} z_n dx$$

$$= \int_{\Omega_{h_0}} |\overline{H}_n \nabla_{h_0} z_n|^2 |\det H_n| dx - \int_{\Omega_{h_0}} (I - \overline{H}_n) \nabla_{h_0} v \cdot H_n^{-1} \nabla_{h_0} z_n |\det H_n| dx$$

$$- \int_{\Omega_{h_0}} \nabla_{h_0} v \cdot (I - \overline{H}_n) \nabla_{h_0} z_n |\det H_n| dx - \int_{\Omega_{h_0}} (1 - |\det H_n|) \nabla_{h_0} v \cdot \nabla_{h_n} z_n dx.$$

Here, I denotes the identity matrix.

We now estimate each term in the equality (5.15). For the left-hand side of the equality (5.15), we have

$$\int_{\Omega_{h_0}} (z_n)_t z_n |\det H_n| dx = \frac{1}{2} \frac{d}{dt} \int_{\Omega_{h_0}} |z_n|^2 |\det H_n| dx, \tag{5.16}$$

and using the Hölder inequality, we get

$$\int_{\Omega_{h_0}} (|\det H_n| - 1) v_t z_n dx \leq \||\det H_n| - 1\|_\infty \left(\|v\|_{H_0^1(\Omega_{h_0})} \|z_n\|_{H_0^1(\Omega_{h_0})} \right.$$

$$+ \|f(v)\|_{L^q(\Omega_{h_0})} \|z_n\|_{L^p(\Omega_{h_0})} \Big)$$

$$+ \||\det H_n| - 1\|_\infty \|g\|_{L^2(\Omega_{h_0})} \|z_n\|_{L^2(\Omega_{h_0})}$$

$$\leq \||\det H_n| - 1\|_\infty \left(\frac{1}{2} \|v\|_{H_0^1(\Omega_{h_0})}^2 + \frac{1}{2} \|z_n\|_{H_0^1(\Omega_{h_0})}^2 \right. \tag{5.17}$$

$$+ \frac{1}{q} \|f(v)\|_{L^q(\Omega_{h_0})}^q + \frac{1}{p} \|z_n\|_{L^p(\Omega_{h_0})}^p \Big)$$

$$+ \||\det H_n| - 1\|_\infty \left(\frac{1}{2} \|g\|_{L^2(\Omega_{h_0})} + \frac{1}{2} \|z_n\|_{L^2(\Omega_{h_0})} \right),$$

where $\| \cdot \|_\infty$ denotes the norm in $L^\infty(\Omega_{h_0})$.

For the right-hand side of Eq. (5.15), the first three terms are estimated as

$$\int_{\Omega_{h_0}} \left(I - \overline{H}_n \right) \nabla_{h_0} v \cdot \overline{H}_n \nabla_{h_0} z_n |\det H_n| dx$$

$$+ \int_{\Omega_{h_0}} \nabla_{h_0} v \cdot (I - \overline{H}_n) \nabla_{h_0} z_n |\det H_n| dx + \int_{\Omega_{h_0}} (1 - |\det H_n|) \nabla_{h_0} v \cdot \nabla_{h_0} z_n dx$$

$$\leq \|I - \overline{H}_n\|_\infty \|\det H_n\|_\infty \|\overline{H}_n\|_\infty \|v\|_{H_0^1(\Omega_{h_0})} \|z_n\|_{H_0^1(\Omega_{h_0})}$$

$$+ \|I - \overline{H}_n\|_\infty \|\det H_n\|_\infty \|v\|_{H_0^1(\Omega_{h_0})} \|z_n\|_{H_0^1(\Omega_{h_0})}$$

$$+ \||\det H_n| - 1\|_\infty \|v\|_{H_0^1(\Omega_{h_0})} \|z_n\|_{H_0^1(\Omega_{h_0})}$$

$$\leq \frac{1}{2} \left\{ \|I - \overline{H}_n\|_\infty \|\det H_n\|_\infty \left(\|\overline{H}_n\|_\infty + 1 \right) + \||\det H_n| - 1\|_\infty \right\} \tag{5.18}$$

$$\cdot \left(\|v\|_{H_0^1(\Omega_{h_0})}^2 + \|z_n\|_{H_0^1(\Omega_{h_0})}^2 \right),$$

where $\|A\|_\infty = \sup\limits_{x\in\overline{\Omega}_{h_0}} \{|a_{ij}(x)|, \ i,j = 1,\ldots,N\}$ for any all $N \times N$ matrix $A = (a_{ij}(x))_{i,j=1}^N$. The fourth term is estimated as

$$\int_{\Omega_{h_0}} (1 - |\det H_n|) f(v) z_n dx \le \|| \det H_n| - 1\|_\infty \|f(v)\|_{L^q(\Omega_{h_0})} \|z_n\|_{L^p(\Omega_{h_0})}$$

$$\le \|| \det H_n| - 1\|_\infty \left(\frac{1}{q} \|f(v)\|_{L^q(\Omega_{h_0})}^q + \frac{1}{p} \|z_n\|_{L^p(\Omega_{h_0})}^p \right),$$
$$(5.19)$$

and the fifth term is estimated as

$$-\int_{\Omega_{h_0}} (f(v_n) - f(v)) z_n |\det H_n| dx \le \ell \int_{\Omega_{h_0}} |z_n|^2 |\det H_n| dx, \qquad (5.20)$$

where the condition (5.3) is used. Finally, the sixth and seventh terms obey the estimates

$$\int_{\Omega_{h_0}} (i_n(g) - g) z_n |\det H_n| dx \le \|i_n(g) - g\|_{L^2(\Omega_{h_0})} \|z_n\|_{L^2(\Omega_{h_0})} \|\det H_n\|_\infty \qquad (5.21)$$

and

$$\int_{\Omega_{h_0}} g z_n (|\det H_n| - 1) dx \le \|g\|_{L^2(\Omega_{h_0})} \|z_n\|_{L^2(\Omega_{h_0})} \|| \det H_n| - 1\|_\infty. \qquad (5.22)$$

By substituting the estimates from (5.16) to (5.22) into (5.15), we deduce that

$$\frac{1}{2}\frac{d}{dt} \int_{\Omega_{h_0}} |z_n|^2 |\det H_n| dx + \int_{\Omega_{h_0}} |\overline{H}_n \nabla_{h_0} z_n|^2 |\det H_n| dx$$

$$\le \ell \int_{\Omega_{h_0}} |z_n|^2 |\det H_n| dx + \|| \det H_n| - 1\|_\infty K(t) \qquad (5.23)$$

$$+ \frac{1}{2} \left\{ \|I - \overline{H}_n\|_\infty \|\det H_n\|_\infty \left(\|\overline{H}_n\|_\infty + 1 \right) + \|| \det H_n| - 1\|_\infty \right\}$$

$$\cdot \left(\|v\|_{H_0^1(\Omega_{h_0})}^2 + \|z_n\|_{H_0^1(\Omega_{h_0})}^2 \right)$$

$$+ \|i_n(g) - g\|_{L^2(\Omega_{h_0})} \|z_n\|_{L^2(\Omega_{h_0})} \|\det H_n\|_\infty,$$

where

$$K(t) = \frac{1}{2}\left(\|v\|^2_{H^1_0(\Omega_{h_0})} + \|z_n\|^2_{H^1_0(\Omega_{h_0})}\right) + 2\left(\frac{1}{q}\|f(v)\|^q_{L^q(\Omega_{h_0})} + \frac{1}{p}\|z_n\|^p_{L^p(\Omega_{h_0})}\right)$$

$$+ \|g\|_{L^2(\Omega_{h_0})} + \|z_n\|_{L^2(\Omega_{h_0})}.$$

This implies that

$$\frac{d}{dt}\left(e^{-2\ell t}\int_{\Omega_{h_0}} |z_n|^2 |\det H_n| dx\right)$$

$$\leq 2e^{-2\ell t}\|\,|\det H_n| - 1\|_\infty K(t) + e^{-2\ell t}\left(\|v\|^2_{H^1_0(\Omega_{h_0})} + \|z_n\|^2_{H^1_0(\Omega_{h_0})}\right) \qquad (5.24)$$

$$\times \left\{\|I - \overline{H}_n\|_\infty\|\det H_n\|_\infty \left(\|\overline{H}_n\|_\infty + 1\right) + \|\,|\det H_n| - 1\|_\infty\right\}$$

$$+ 2e^{-2\ell t}\|i_n(g) - g\|_{L^2(\Omega_{h_0})}\|z_n\|_{L^2(\Omega_{h_0})}\|\det H_n\|_\infty.$$

By integrating (5.24) from 0 to $t \in [0, T]$, we derive that

$$\int_{\Omega_{h_0}} |z_n|^2 |\det H_n| dx$$

$$\leq e^{2\ell t}\int_{\Omega_{h_0}} |z_n(0)|^2 |\det H_n| dx + 2e^{2\ell t}\|\,|\det H_n| - 1\|_\infty \int_0^T K(t) dt \qquad (5.25)$$

$$+ e^{2\ell t}\left\{\|I - \overline{H}_n\|_\infty\|\det H_n\|_\infty \left(\|\overline{H}_n\|_\infty + 1\right) + \|\,|\det H_n| - 1\|_\infty\right\}$$

$$\times \int_0^T \left(\|v(t)\|^2_{H^1_0(\Omega_{h_0})} + \|z_n(t)\|^2_{H^1_0(\Omega_{h_0})}\right) dt$$

$$+ \frac{e^{2\ell t}}{\ell}\|\det H_n\|_\infty\|z_n\|_{L^\infty(0,T;L^2(\Omega_{h_0}))}\|i_n(g) - g\|_{L^2(\Omega_{h_0})}.$$

The estimates (5.6), (5.7), and (5.8) imply that there is a constant $C > 0$ (independent of n) such that

$$\int_0^T K(t) dt \leq C, \quad \|z_n\|_{L^\infty(0,T;L^2(\Omega_{h_0}))} \leq C,$$

$$\int_0^T \left(\|v(t)\|^2_{H^1_0(\Omega_{h_0})} + \|z_n(t)\|^2_{H^1_0(\Omega_{h_0})}\right) dt \leq C.$$

Since $h_n \to h_0$ in $\mathrm{Diff}(\Omega_0)$ and det is a continuous operator, $\|\det H_n\|_\infty$ and $\|\overline{H}_n\|_\infty$ are uniformly bounded with respect to n. Therefore, from (5.25), we conclude that

$$\int_{\Omega_{h_0}} |z_n|^2 |\det H_n| dx \leq e^{2\ell t} \int_{\Omega_{h_0}} |z_n(0)|^2 |\det H_n| dx + Ce^{2\ell t} \|i_n(g) - g\|_{L^2(\Omega_{h_0})}$$

$$+ Ce^{2\ell t} \left(\| |\det H_n| - 1 \|_\infty + \| I - \overline{H}_n \|_\infty \right).$$

Finally, we deduce that

$$\inf_{x \in \Omega_{h_0}} |\det H_n(x)| \|z_n(t)\|^2_{L^2(\Omega_{h_0})}$$

$$\leq e^{2\ell t} \|\det H_n\|_\infty \|z_n(0)\|^2_{L^2(\Omega_{h_0})} \tag{5.26}$$

$$+ Ce^{2\ell t} \left(\|i_n(g) - g\|_{L^2(\Omega_{h_0})} + \| |\det H_n| - 1 \|_\infty + \| I - \overline{H}_n \|_\infty \right).$$

Note that $\inf\limits_{x \in \Omega_{h_0}} |\det H_n(x)| \geq \frac{1}{2}$ for all sufficiently large $n \in \mathbb{N}$. Moreover, we get

$$\| |\det H_n| - 1 \|_\infty \to 0 \text{ and } \| I - \overline{H}_n \|_\infty \to 0 \text{ as } n \to \infty.$$

Since $C_c^\infty(\mathbb{R}^N)$ is dense in $L^2(\mathbb{R}^N)$ and $g \in L^2(\mathbb{R}^N)$, we can choose $\xi_m \in C_c^\infty(\mathbb{R}^N)$ such that $\xi_m \to g$ in $L^2(\mathbb{R}^N)$. Hence, we have

$$\|i_n(g) - g\|_{L^2(\Omega_{h_0})} \leq \|\det H_n\|_\infty \|g - \xi_m\|_{L^2(\mathbb{R}^N)} + \|i_n(\xi_m) - \xi_m\|_{L^2(\Omega_{h_0})}$$

$$+ \|\xi_m - g\|_{L^2(\Omega_{h_0})}.$$

This implies that $\|i_n(g) - g\|_{L^2(\Omega_{h_0})} \to 0$ as $n \to \infty$, and now completes (5.26) the proof of the lemma. □

With the above lemma and proposition at hand, we are ready to prove Theorem 5.1.

End of Proof of Theorem 5.1 We prove the theorem by showing that the assumptions of Proposition 2.1 are satisfied. To this and, we let $\Lambda = \text{Diff}(\Omega)$ and $D_h = B(0, R_h)$ for each $h \in \text{Diff}(\Omega)$, where R_h is a constant given in (5.9). Then it is enough to show that for any $T > 0$, $\varepsilon > 0$, and $h_0 \in \text{Diff}(\Omega)$, there exists a neighborhood $\mathcal{W}$ of h_0 such that for any $h \in \mathcal{W}$, there is an ε-immersion $i : D_h \to L^2(\Omega_{h_0})$ satisfying

$$i(D_h) \subset B(D_{h_0}, \varepsilon), \text{ and}$$

$$\|i(S_h(u_0, t)) - S_{h_0}(i(u_0), t)\|_{L^2(\Omega_{h_0})} < \varepsilon \tag{5.27}$$

for all $u_0 \in \mathcal{A}_h$ and $t \in [0, T]$. Suppose this is not the case. Then there exist $T > 0$, $\varepsilon > 0$, and a sequence $\{h_n\}_{n \in \mathbb{N}}$ converging to h_0 in $\text{Diff}(\Omega)$ such that for any ε-immersion $i_n : D_{h_n} \to L^2(\Omega_{h_0})$ satisfying $i_n(D_{h_n}) \subset B(D_{h_0}, \varepsilon)$, there are $u_{n,0} \in \mathcal{A}_{h_n}$ and $t_n \in [0, T]$ such that

$$\|i_n(S_h(u_{n,0}, t_n)) - S_{h_0}(i_n(u_{n,0}), t_n)\|_{L^2(\Omega_{h_0})} > \varepsilon. \tag{5.28}$$

In particular, take $i_n = i_{h_n}$, where i_{h_n} is given as before, that is, $i_{h_n}(u) = u \circ (h_n \circ h_0^{-1})$ for $u \in L^2(\Omega_{h_n})$. Let $v_{n,0} = i_n(u_{n,0})$ for each $n \in \mathbb{N}$. Since $u_{n,0} \in H_0^1(\Omega_{h_n})$ and $\|i_n\| \to 1$ as $n \to \infty$, $\{v_{n,0}\}$ is uniformly bounded in $H_0^1(\Omega_{h_0})$. By the compactness of the embedding $H_0^1(\Omega_{h_0}) \hookrightarrow L^2(\Omega_{h_0})$, $\{v_{n,0}\}$ has a convergent subsequence in $L^2(\Omega_{h_0})$, say

$$v_{n,0} \to v_0 \in L^2(\Omega_{h_0}) \text{ as } n \to \infty.$$

Then (5.28) can be rewritten as

$$\|i_n(S_{h_n}(i_n^{-1}(v_{n,0}), t_n)) - S_{h_0}(v_0, t_n)\|_{L^2(\Omega_{h_0})} \geq \frac{\varepsilon}{2}$$

for all sufficiently large n. This is a contradiction by Lemma 5.2, and so the proof of Theorem 5.1 is complete. $\square$

5.3   Proof of Theorem 5.2

In that follows, we prove that, residually, the dynamical system on the global attractor induced by the problem (5.1) varies continuously in the topology of Gromov-Hausdorff convergence with respect to perturbations of the domain (Theorem 5.2). As a consequence, we can derive the main result (Theorem 2.9) for the semidynamical systems S_h induced by the equations in (5.4) residually with respect to h.

Lemma 5.3 *For any $h_0 \in \text{Diff}(\Omega)$, there is a neighborhood $\mathcal{U}$ of h_0 such that $\{S_h\}_{h \in \mathcal{U}}$ is equicontinuous on $A_h \times [0, 1]$, that is, for any $\varepsilon > 0$, there exists $\delta > 0$ such that if $\|u_0 - \tilde{u}_0\|_{L^2(\Omega_h)} < \delta$ and $|t - s| < \delta$, then $\|S_h(u_0, t) - S_h(\tilde{u}_0, s)\|_{L^2(\Omega_h)} < \varepsilon$ for all $h \in \mathcal{U}$, $u_0, \tilde{u}_0 \in A_h$, and $t, s \in [0, 1]$.*

Proof Step 1. We show that for any $\varepsilon > 0$ and $h \in \text{Diff}(\Omega)$, there is $\delta_1 > 0$ such that if $\|u_0 - \tilde{u}_0\|_{L^2(\Omega_h)} < \delta_1$, then

$$\|S_h(u_0, t) - S_h(\tilde{u}_0, t)\|_{L^2(\Omega_h)} < \frac{\varepsilon}{2} \tag{5.29}$$

for any $t \in [0, 1]$. Let $\ell > 0$ be a constant satisfying (5.3), and take a constant δ_1 with $\delta_1 < \frac{\varepsilon}{2e^\ell}$. For any two points $u_0, \tilde{u}_0 \in L^2(\Omega_h)$ with $\|u_0 - \tilde{u}_0\|_{L^2(\Omega_h)} < \delta_1$, we let

$$u(t) = S_h(u_0, t), \quad \tilde{u}(t) = S_h(\tilde{u}_0, t) \text{ and } z(t) = u(t) - \tilde{u}(t), \text{ for all } t \in \mathbb{R}_0^+.$$

Then z is a weak solution of

$$\begin{cases} \partial_t z - \Delta_h z + f(u) - f(\tilde{u}) = 0 & \text{in } \Omega_h \times (0, \infty), \\ z = 0 & \text{on } \partial\Omega_h \times (0, \infty), \\ z(x, 0) = z_0 := u_0 - \tilde{u}_0 & \text{in } \Omega_h. \end{cases}$$

Hence, we have

$$\frac{1}{2}\frac{d}{dt}\|z\|^2_{L^2(\Omega_h)} + \|\nabla_h z\|^2_{L^2(\Omega_h)} = -\int_{\Omega_h} (f(u) - f(\tilde{u}))z \, dx.$$

Using (5.3), we obtain the estimate

$$\frac{d}{dt}\|z\|^2_{L^2(\Omega_h)} + 2\|\nabla_h z\|^2_{L^2(\Omega_h)} \leq 2\ell\|z\|^2_{L^2(\Omega_h)}.$$

In particular,

$$\frac{d}{dt}\|z\|^2_{L^2(\Omega_h)} \leq 2\ell\|z\|^2_{L^2(\Omega_h)}.$$

By the Gronwall inequality, we have

$$\|z\|^2_{L^2(\Omega_h)} \leq e^{2\ell t}\|z_0\|^2_{L^2(\Omega_h)} \leq e^{2\ell}\|u_0 - \tilde{u}_0\|^2_{L^2(\Omega_h)} < \frac{\varepsilon^2}{4}, \quad \text{for all } t \in [0, 1].$$

Step 2. We claim that for any $h_0 \in \text{Diff}(\Omega)$ and $\varepsilon > 0$, there exist a neighborhood $\mathcal{U}$ of h_0 and $\delta_2 > 0$ such that for any $h \in \mathcal{U}$,

$$\|S_h(u_0, s) - u_0\|_{L^2(\Omega_h)} < \frac{\varepsilon}{2} \tag{5.30}$$

for all $(u_0, s) \in \mathcal{A}_h \times [0, \delta_2]$. As in the proof of Theorem 5.1, we can take a neighborhood $\mathcal{U}$ of h_0 such that for any $h \in \mathcal{U}$, there is an $\frac{\varepsilon}{6}$-immersion $i_h : B_h \to L^2(\Omega_{h_0})$ satisfying

$$\|i_h(S_h(u_0, s)) - S_{h_0}(i_h(u_0), s)\|_{L^2(\Omega_{h_0})} < \frac{\varepsilon}{6}$$

for all $(h, s) \in \mathcal{U} \times [0, 1]$. Since B_{h_0} is compact in $L^2(\Omega_{h_0})$ and $i_h(u_0) \in B_{h_0}$ for any $u_0 \in \mathcal{A}_h$, there exists $\delta_2 > 0$ such that

$$\|S_{h_0}(i_h(u_0), s) - i_h(u_0)\|_{L^2(\Omega_{h_0})} < \frac{\varepsilon}{6} \tag{5.31}$$

for all $(u_0, s) \in A_h \times [0, \delta_2]$. Suppose this is not the case. Then for each $n \in \mathbb{N}$, there are $s_n \in [0, 1/n]$ and $u_{n,0} \in A_h$ satisfying

$$\|S_{h_0}(i_h(u_{n,0}), s_n) - i_h(u_{n,0})\|_{L^2(\Omega_{h_0})} \geq \frac{\varepsilon}{6}.$$

Since B_{h_0} is compact in $L^2(\Omega_{h_0})$, we may assume that as n tends to infinity $i_h(u_{n,0})$ converges in $L^2(\Omega_{h_0})$ to a point, say $v_0 \in B_{h_0}$. Since

$$\|S_{h_0}(v_0, s_n) - S_{h_0}(i_h(u_{n,0}), s_n)\|_{L^2(\Omega_{h_0})} < e^{\ell}\|v_0 - i_h(u_{n,0})\|_{L^2(\Omega_{h_0})},$$

there exists $N_1 > 0$ such that for all $n > N_1$,

$$\max\left\{\|S_{h_0}(v_0, s_n) - S_{h_0}(i_h(u_{n,0}), s_n)\|_{L^2(\Omega_{h_0})}, \|i_h(u_{n,0}) - v_0\|_{L^2(\Omega_{h_0})}\right\} < \frac{\varepsilon}{18}.$$

Note that $s_n \to 0$ as $n \to \infty$. By the continuity of $S_{h_0}(v_0, \cdot)$, we can choose $N_2 > 0$ such that

$$\|S_{h_0}(v_0, s_n) - v_0\|_{L^2(\Omega_{h_0})} < \frac{\varepsilon}{18}$$

for all $n > N_2$. Then for any $n > \max\{N_1, N_2\}$, we have

$$\begin{aligned}
0 = \|S_{h_0}(v_0, 0) - v_0\| \geq \ &\|S_{h_0}(i_h(u_{n,0}), s_n) - i_h(u_{n,0})\|_{L^2(\Omega_{h_0})} \\
&- \|S_{h_0}(i_h(u_{n,0}), s_n) - S_{h_0}(v_0, s_n)\|_{L^2(\Omega_{h_0})} \\
&- \|S_{h_0}(v_0, s_n) - S_{h_0}(v_0, 0)\|_{L^2(\Omega_{h_0})} \\
&- \|i_h(u_{n,0}) - v_0\|_{L^2(\Omega_{h_0})} > 0.
\end{aligned}$$

This contradiction proves (5.31).

It follows that for any $h \in \mathcal{U}$, $u_0 \in A_h$, and $s \in [0, \delta_2]$,

$$\begin{aligned}
\|S_h(u_0, s) - u_0\|_{L^2(\Omega_h)} \leq \ &\frac{\varepsilon}{6} + \|i_h(S_h(u_0, s)) - i_h(u_0)\|_{L^2(\Omega_{h_0})} \\
\leq \ &\frac{\varepsilon}{6} + \|i_h(S_h(u_0, s)) - S_{h_0}(i_h(u_0), s)\|_{L^2(\Omega_{h_0})} \\
&+ \|S_{h_0}(i_h(u_0), s) - i_h(u_0)\|_{L^2(\Omega_{h_0})} \leq \frac{\varepsilon}{2}.
\end{aligned}$$

Step 3. For any $h_0 \in \text{Diff}(\Omega)$ and $\varepsilon > 0$, let $\mathcal{U}$ be a neighborhood of h_0 as in Step 2 and let $\delta = \min\{\delta_1, \delta_2\}$, where δ_1 and δ_2 are taken from Steps 1 and 2, respectively. Let $h \in \mathcal{U}$, $u_0, \tilde{u}_0 \in A_h$, and $t, s \in [0, 1]$ be such that $\|u_0 - \tilde{u}_0\|_{L^2(\Omega_h)} < \delta$ and $|t - s| < \delta$. A applying inequalities (5.29) and (5.30), we have

$$\|S_h(u_0, t) - S_h(\tilde{u}_0, s)\|_{L^2(\Omega_h)} \leq \|S_h(u_0, t) - S_h(\tilde{u}_0, t)\|_{L^2(\Omega_h)}$$

$$+ \|S_h(u_0, t) - S_h(\tilde{u}_0, s)\|_{L^2(\Omega_h)}$$

$$\leq \frac{\varepsilon}{2} + \|S_h(S_h(u_0, s), t - s) - S_h(\tilde{u}_0, s)\|_{L^2(\Omega_h)}$$

$$\leq \varepsilon.$$

This completes the proof of Lemma 5.3.   □

Remark 5.1 Lemma 5.3 implies that for any $h_0 \in \mathrm{Diff}(\Omega)$ and $T > 0$, there is a neighborhood $\mathcal{U}$ of h_0 such that $\{S_h\}_{h \in \mathcal{U}}$ is equicontinuous on $\mathcal{A}_h \times [0, T]$.

End of Proof of Theorem 5.2 For any $h, k \in \mathrm{Diff}(\Omega)$, we define an isomorphism i_{hk} : $L^2(\Omega_h) \to L^2(\Omega_k)$ by $i_{hk}(u) = u \circ (h \circ k^{-1})$. We observe that for any $\varepsilon > 0$, there is $\delta > 0$ such that if $d_{C^1}(h, k) < \delta$, then i_{hk} and i_{kh} are ε-immersions, where d_{C^1} denotes the C^1-metric on $\mathrm{Diff}(\Omega)$.

Step 1. We prove that there is a residual subset $\mathcal{R}$ of $\mathrm{Diff}(\Omega)$ such that for any $k \in \mathcal{R}$ and $\varepsilon > 0$, there is a neighborhood $\mathcal{W}$ of k such that for any $h \in \mathcal{W}$, the ε-immersion i_{hk} satisfies

$$i_{hk}(\mathcal{A}_h) \subset B(\mathcal{A}_k, \varepsilon) \quad \text{and} \quad \mathcal{A}_k \subset B(i_{hk}(\mathcal{A}_h), \varepsilon).$$

For any $\varepsilon > 0$, we denote by $F(\varepsilon)$ the collection of all $k \in \mathrm{Diff}(\Omega)$ such that there is a neighborhood $\mathcal{W}$ of k with the following property: for any $h, \tilde{h} \in \mathcal{W}$, the ε-immersion $i_{h\tilde{h}}$ satisfies $i_{h\tilde{h}}(\mathcal{A}_h) \subset B(\mathcal{A}_{\tilde{h}}, \varepsilon)$. It is clear that $F(\varepsilon)$ is open in $\mathrm{Diff}(\Omega)$ for any $\varepsilon > 0$. Let $\mathcal{R} = \bigcap_{n \in \mathbb{N}} F(1/n)$. To prove that $\mathcal{R}$ is a residual subset of $\mathrm{Diff}(\Omega)$, it is sufficient to show that $F(8\varepsilon)$ is dense in $\mathrm{Diff}(\Omega)$.

Let $\mathcal{U}$ be a nonempty open set in $\mathrm{Diff}(\Omega)$, and $h_0 \in \mathcal{U}$. Let $\{B_i = B(u_i, \varepsilon/3), u_i \in L^2(\Omega_{h_0})\}_{i \in K}$ be a finite open cover of $\mathcal{A}_{h_0}$ such that $\mathcal{A}_{h_0} \cap B_i \neq \emptyset$ for any $i \in K$. Let $\mathcal{I}$ be the collection of subsets $J \subset K$ such that there are $k \in \mathcal{U}$ and an injective δ-immersion $j_k : D_k \to L^2(\Omega_{h_0})$ for some $\delta < \varepsilon/2$ satisfying

$$j_k(\mathcal{A}_k) \cap B_i \neq \emptyset \quad \text{for all } i \in J, \text{ and}$$

$$j_k(U_k) \subset \bigcup_{i \in J} B_i \quad \text{for a neighborhood } U_k \text{ of } \mathcal{A}_k. \tag{5.32}$$

We note that $\mathcal{I}$ is nonempty since it contains K. Choose a minimal element J of $\mathcal{I}$. Then, by the definition of $\mathcal{I}$, there are $k \in \mathcal{U}$ and an injective δ-immersion $j_k : D_k \to L^2(\Omega_{h_0})$ for some $\delta < \varepsilon/2$ satisfying (5.32). By Step 1 in Proposition 2.1, there is a neighborhood $\mathcal{W} \subset \mathcal{U}$ of k such that for any $h \in \mathcal{W}$, i_{hk} an the $(\varepsilon/4 - \delta/2)$-immersion satisfying

$$\|i_{kh}\| < 2, \quad \text{and } i_{hk}(U_h) \subset U_k \text{ for a neighborhood } U_h \text{ of } \mathcal{A}_h.$$

We can check that $j_k \circ i_{hk}$ is an injective $(\varepsilon/4 + \delta/2)$-immersion from D_h to $L^2(\Omega_{h_0})$ such that

$$j_k \circ i_{hk}(U_h) \subset j_k(U_k) \subset \bigcup_{i \in J} B_i.$$

Since J is a minimal element in $\mathcal{I}$, we have $j_k \circ i_{hk}(A_h) \cap B_i \neq \emptyset$ for all $i \in J$. We observe that

$$A_k \subset \bigcup_{i \in J} B(j_k^{-1}(u_i), \varepsilon), \quad A_k \cap B(j_k^{-1}(u_i), \varepsilon) \neq \emptyset,$$

$$i_{hk}(A_h) \subset \bigcup_{i \in J} B(j_k^{-1}(u_i), \varepsilon), \text{ and } i_{hk}(A_h) \cap B(j_k^{-1}(u_i), \varepsilon) \neq \emptyset$$

for all $i \in J$. Therefore,

$$A_k \subset B(i_{hk}(A_h), 2\varepsilon) \text{ and } i_{hk}(A_h) \subset B(A_k, 2\varepsilon)$$

for all $h \in \mathcal{W}$, and so

$$i_{hk}(A_h) \subset B(i_{\tilde{h}k}(A_{\tilde{h}}), 4\varepsilon)$$

for all $h, \tilde{h} \in \mathcal{W}$. We conclude that

$$i_{h\tilde{h}}(A_h) = i_{k\tilde{h}} \circ i_{hk}(A_h) \subset i_{k\tilde{h}}(B(i_{\tilde{h}k}(A_{\tilde{h}}), 4\varepsilon)) \subset B(A_{\tilde{h}}, 8\varepsilon).$$

Consequently, we have proved $k \in \mathcal{U} \cap F(8\varepsilon)$. So, $F(8\varepsilon)$ is dense in $\mathrm{Diff}(\Omega)$.

Finally, for any $k \in \mathcal{R}$ and $\varepsilon > 0$, choose $n \in \mathbb{N}$ with $1/n < \varepsilon/2$. Since $k \in F(1/n)$, there is a neighborhood $\mathcal{W}$ of k such that for any $h, \tilde{h} \in \mathcal{W}$, the $(1/n)$-immersion $i_{h\tilde{h}}$ satisfies $i_{h\tilde{h}}(A_h) \subset B(A_h, 1/n)$. In particular, letting $k = \tilde{h}$, we obtain

$$i_{hk}(A_h) \subset B(A_k, \varepsilon) \text{ and } A_k \subset B(i_{hk}(A_h), \varepsilon)$$

for all $h \in \mathcal{W}$.

Step 2. We prove that the map $\phi : \mathrm{Diff}(\Omega) \to \mathcal{CDS}$ defined by $\phi(h) = \phi_h$ is continuous on $\mathcal{R}$.

Fix $k \in \mathcal{R}$ and $\varepsilon > 0$. By Lemma 5.2, there exists $0 < \delta < \varepsilon/3$ such that if $\|u - \hat{u}\|_{L^2(\Omega_k)} < \delta$ for $u \in L^2(\Omega_k)$ and $\hat{u} \in A_k$, then

$$\|S_k(u, t) - S_k(\hat{u}, t)\|_{L^2(\Omega_k)} < \frac{\varepsilon}{3}$$

for $t \in [0, 1]$. By Step 1 and (5.27), we can choose a neighborhood $\mathcal{W}$ of k in $\text{Diff}(\Omega)$ such that for any $h \in \mathcal{W}$, i_{hk} is the δ-immersion i_{hk} satisfies

$$i_{hk}(\mathcal{A}_h) \subset B(\mathcal{A}_k, \delta), \qquad \mathcal{A}_k \subset B(i_{hk}(\mathcal{A}_h), \delta), \tag{5.33}$$

$$\|i_{hk}(S_h(u, t)) - S_k(i_{hk}(u), t)\|_{L^2(\Omega_h)} < \delta$$

for all $(u, t) \in \mathcal{A}_h \times [0, 1]$. For any $h \in \mathcal{W}$, we take a map $\hat{i}_{hk} : \mathcal{A}_h \to \mathcal{A}_k$ satisfying

$$\hat{i}_{hk}(u) \in \mathcal{A}_k \quad \text{for } u \in \mathcal{A}_h \text{ and } \|\hat{i}_{hk}(u) - i_{hk}(u)\|_{L^2(\Omega_k)} < \delta.$$

Then $\hat{i}_{hk}$ is an ε-isometry. Indeed, for any $u, \hat{u} \in A_h$, we have

$$\|\hat{i}_{hk}(u) - \hat{i}_{hk}(\hat{u})\|_{L^2(\Omega_k)} - \|u - \hat{u}\|_{L^2(\Omega_k)}$$

$$\leq \|\hat{i}_{hk}(u) - i_{hk}(u)\|_{L^2(\Omega_k)} + \|i_{hk}(\hat{u}) - \hat{i}_{hk}(\hat{u})\|_{L^2(\Omega_k)}$$

$$+ \|i_{hk}(u) - i_{hk}(\hat{u})\|_{L^2(\Omega_k)} - \|u - \hat{u}\|_{L^2(\Omega_k)} < \varepsilon.$$

Similarly, we can show that

$$\|u - \hat{u}\|_{L^2(\Omega_k)} - \|\hat{i}_{hk}(u) - \hat{i}_{hk}(\hat{u})\|_{L^2(\Omega_k)} < \varepsilon.$$

By (5.33), for any $u \in \mathcal{A}_k$, we can choose $\tilde{u} \in \mathcal{A}_h$ such that $\|i_{hk}(\tilde{u}) - u\|_{L^2(\Omega_k)} < \delta$. Since

$$\|\hat{i}_{hk}(\tilde{u}) - u\|_{L^2(\Omega_k)} \leq \|\hat{i}_{hk}(\tilde{u}) - i_{hk}(\tilde{u})\|_{L^2(\Omega_k)} + \|i_{hk}(\tilde{u}) - u\|_{L^2(\Omega_k)} < \varepsilon,$$

we get $\mathcal{A}_k = B(\hat{i}_{hk}(\mathcal{A}_h), \varepsilon)$, so $\hat{i}_{hk}$ is an ε-isometry, as claimed.

Moreover, we note that

$$\|\hat{i}_{hk}(\phi_h(u, t)) - \phi_k(\hat{i}_{hk}(u), t)\|_{L^2(\Omega_k)} \leq \|\hat{i}_{hk}(\phi_h(u, t)) - i_{hk}(\phi_h(u, t))\|_{L^2(\Omega_k)}$$

$$+ \|i_{hk}(\phi_h(u, t)) - \phi_k(i_{hk}(u), t)\|_{L^2(\Omega_k)}$$

$$+ \|\phi_k(i_{hk}(u), t) - \phi_k(\hat{i}_{hk}(u), t)\|_{L^2(\Omega_k)}$$

$$< \varepsilon \tag{5.34}$$

for all $(u, t) \in \mathcal{A}_h \times [0, 1]$. Let $\{h_n\}_{n \in \mathbb{N}}$ be a sequence in $\text{Diff}(\Omega)$ converging to $k \in \mathcal{R}$. By (5.34) and Remark 2.1, for each $n \in \mathbb{N}$ there are $(1/n)$-isometries $\hat{i}_n : \mathcal{A}_{h_n} \to \mathcal{A}_k$ and $\hat{j}_n : \mathcal{A}_k \to \mathcal{A}_{h_n}$ such that

$$d_{C^0}(\hat{j}_n \circ \hat{i}_n, id_{\mathcal{A}_{h_n}}) < \frac{1}{n}, \ d_{C^0}(\hat{i}_n \circ \hat{j}_n, id_{\mathcal{A}_k}) < \frac{1}{n}$$

and

$$\|\hat{i}_n(\phi_{h_n}(u,t)) - \phi_k(\hat{i}_n(u),t)\|_{L^2(\Omega_k)} < \frac{1}{n}$$

for all $(u,t) \in \mathcal{A}_{h_n} \times [0,1]$. Let A be a countable dense subset of $\mathcal{A}_k$. By the diagonal argument, we may assume that for each $u \in A$ and $t \in [0,1] \cap \mathbb{Q}$, the sequence $\{\hat{i}_n(S_{h_n}(\hat{j}_n(u),t))\}_{n\in\mathbb{N}}$ converges to a point, say $\phi_*(u,t)$, in $\mathcal{A}_k$. Hence, we can define a map $\phi_* : A \times ([0,1] \cap \mathbb{Q}) \to \mathcal{A}_k$ by

$$\phi_*(u,t) = \lim_{n\to\infty} \hat{i}_n(\phi_{h_n}(\hat{j}_n(u),t))$$

for $(u,t) \in A \times ([0,1] \cap \mathbb{Q})$. We claim that ϕ_* is uniformly continuous. To show this, for any $\varepsilon > 0$, take a constant $\delta > 0$ corresponding to $\varepsilon/2$ by Lemma 5.3. Then for any $(u,t), (\hat{u},\hat{t}) \in A \times ([0,1] \cap \mathbb{Q})$ with $\|u - \hat{u}\|_{L^2(\Omega_k)} + |t - \hat{t}| < \delta$, we have

$$\|\phi_*(u,t) - \phi_*(\hat{u},\hat{t})\|_{L^2(\Omega_k)} = \lim_{n\to\infty} \|\hat{i}_n(\phi_{h_n}(\hat{j}_n(u),t)) - \hat{i}_n(\phi_{h_n}(\hat{j}_n(\hat{u}),\hat{t}))\|_{L^2(\Omega_k)}$$

$$\leq \lim_{n\to\infty} \left(\frac{1}{n} + \|\phi_{h_n}(\hat{j}_n(u),t) - \phi_{h_n}(\hat{j}_n(\hat{u}),\hat{t})\|_{L^2(\Omega_k)} \right)$$

$$< \varepsilon,$$

as needed. Now the uniformly continuous map ϕ_* of $A \times ([0,1] \cap \mathbb{Q})$ into $\mathcal{A}_k$ can be extended to a uniformly continuous map (still denoted by ϕ_*) of $\mathcal{A}_k \times [0,1]$ into $\mathcal{A}_k$.

For any $(u,t) \in A \times ([0,1] \cap \mathbb{Q})$, we have

$$\|\phi_*(u,t) - \phi_k(u,t)\|_{L^2(\Omega_k)} = \lim_{n\to\infty} \|\hat{i}_n(\phi_{h_n}(\hat{j}_n(u),t)) - \phi_k(u,t)\|_{L^2(\Omega_k)}$$

$$\leq \lim_{n\to\infty} \|\hat{i}_n(\phi_{h_n}(\hat{j}_n(u),t)) - \phi_k(\hat{i}_n \circ \hat{j}_n(u),t)\|_{L^2(\Omega_k)}$$

$$+ \|\phi_k(\hat{i}_n \circ \hat{j}_n(u),t) - \phi_k(u,t)\|_{L^2(\Omega_k)}$$

$$\leq \lim_{n\to\infty} \left(\frac{1}{n} + \|\phi_k(\hat{i}_n \circ \hat{j}_n(u),t) - \phi_k(u,t)\|_{L^2(\Omega_k)} \right)$$

$$= 0.$$

This implies that $\phi_* = \phi_k$ on $\mathcal{A}_k \times [0,1]$.

Next, let us show that if $(u,t) \in \mathcal{A}_k \times [0,1]$, then the sequence $\{\hat{i}_n(\phi_{h_n}(\hat{j}_n(u),t))\}_{n\in\mathbb{N}}$ converges to $\phi_k(u,t)$. In fact, for any $(u,t) \in \mathcal{A}_k \times [0,1]$, we can take a sequence $\{(u_l,t_l)\}_{l\in\mathbb{N}}$ in $A \times ([0,1] \cap \mathbb{Q})$ converging to (u,t). For any $\varepsilon > 0$, choose a constant $\delta > 0$ $(\delta < \varepsilon)$ corresponding to $\varepsilon/4$ by Lemma 5.3. Choose $l \in \mathbb{N}$ satisfying

$$|t_l - t| \leq \frac{\delta}{4} \quad \text{and} \quad \|u_l - u\|_{L^2(\Omega_k)} \leq \frac{\delta}{4}.$$

By the definition of $\hat{i}_n$ and the continuity of ϕ_k, there exists $N_3 > 0$ such that $1/N_3 < \delta/4$ and

$$\|\hat{i}_n(\phi_{h_n}(\hat{j}_n(u_l), t_l)) - \phi_k(u_l, t_l)\|_{L^2(\Omega_k)} < \frac{\varepsilon}{4}$$

for all $n > N_3$. Since $\|\hat{j}_n(u_l) - \hat{j}_n(u)\|_{L^2(\Omega_{h_n})} < 1/n + \|u_l - u\|_{L^2(\Omega_k)} < \delta/2$ for all $n > N_3$, we have

$$\|\phi_{h_n}(\hat{j}_n(u_l), t_l) - \phi_{h_n}(\hat{j}_n(u), t)\|_{L^2(\Omega_k)} < \frac{\varepsilon}{4}.$$

It follows that

$$\|\hat{i}_n(\phi_{h_n}(\hat{j}_n(u), t)) - \phi_k(u, t)\|_{L^2(\Omega_k)}$$
$$\leq \|\hat{i}_n(\phi_{h_n}(\hat{j}_n(u), t)) - \hat{i}_n(\phi_{h_n}(\hat{j}_n(u_l), t_l))\|_{L^2(\Omega_k)}$$
$$+ \|\hat{i}_n(\phi_{h_n}(\hat{j}_n(u_l), t_l)) - \phi_k(u_l, t_l)\|_{L^2(\Omega_k)}$$
$$+ \|\phi_k(u_l, t_l) - \phi_k(u, t)\|_{L^2(\Omega_k)} \leq \varepsilon$$

for all $n > N_3$. Thus, we have checked assumptions (1), (2), and (3) of Theorem 2.9 for the semidynamical system S_h induced by the reaction-diffusion equations (5.4) residually. This completes the proof of Theorem 5.2.   □

Remark 5.2 Let $\mathcal{A}_h$ be the global attractor of the semidynamical system S_h induced by Eq. (5.4). Then one can show that $\mathcal{A}_h$ is GH-stable under perturbations of the domain if and only if, for any $T > 0$ and $\varepsilon > 0$, there is $\delta > 0$ such that if $d_{C^1}(h, \tilde{h}) < \delta$, then for any $u \in \mathcal{A}_h$, $\tilde{u} \in \mathcal{A}_{\tilde{h}}$ and $t \in [0, T]$,

$$\|\hat{i}_{h\tilde{h}}(\phi_h(u, t) - \phi_{\tilde{h}}(\hat{i}_{h\tilde{h}}(u), t))\|_{L^2(\Omega_{\tilde{h}})} < \varepsilon, \text{ and}$$

$$\|\hat{j}_{h\tilde{h}}(\phi_{\tilde{h}}(\tilde{u}, t) - \phi_h(\hat{j}_{h\tilde{h}}(\tilde{u}), t))\|_{L^2(\Omega_h)} < \varepsilon,$$

where $\hat{i}_{h\tilde{h}}$ and $\hat{j}_{h\tilde{h}}$ are defined as before.

We say that an equilibrium u_0 of the equation (5.4) is hyperbolic if 0 is not in the spectrum of the linear operator $\Delta - F'(u_0)I : L^2(\Omega_h) \to L^2(\Omega_h)$, where I denotes the identity map on $L^2(\Omega_h)$ and F' is the Nemytskii operator of f'.

Remark 5.3 In this direction, one we can prove the continuity of global attractors and the Gromov-Hausdorff stability of reaction-diffusion equations with Neumann or Robin boundary conditions under perturbations of the domain and equation if every equilibrium of the unperturbed equation is hyperbolic. For more details, see [44, 48].

Stability of Inertial Manifolds

6

6.1 Introduction

In this chapter, we study the Gromov-Hausdorff stability and the continuous dependence of the inertial manifolds under perturbations of the domain and equation. More precisely, we use the Gromov-Hausdorff distances between two inertial manifolds and two dynamical systems to consider the continuous dependence of the inertial manifolds and the stability of the dynamical systems on inertial manifolds induced by reaction-diffusion equations under perturbations of the domain and equation.

Let Ω_0 be an open bounded domain in $\mathbb{R}^N$ with smooth boundary. We consider the dissipativity reaction-diffusion equation

$$\begin{cases} \partial_t u - \Delta u = f_0(u) & \text{in } \Omega_0 \times (0, \infty), \\ u = 0 & \text{on } \partial\Omega_0 \times (0, \infty), \end{cases} \tag{6.1}$$

where $f_0 : \mathbb{R} \to \mathbb{R}$ is a C^1 function such that f_0 and f_0' are bounded, and f_0 satisfies the dissipativity condition,

$$\limsup_{|s| \to \infty} \frac{f_0(s)}{s} < 0.$$

It shown in [7] that the problem (6.1) is well posed in various function spaces. Let $F_0 : L^2(\Omega_0) \to L^2(\Omega_0)$ be the Nemytskii operator generated by f_0. It is clear that F_0 is Lipschitz since f_0' is bounded, and we may assume $\mathrm{Lip}\, F_0 > 1$.

© The Author(s), under exclusive license to Springer Nature Switzerland AG 2022   111
J. Lee, C. Morales, *Gromov-Hausdorff Stability of Dynamical Systems and Applications to PDEs*, Frontiers in Mathematics, https://doi.org/10.1007/978-3-031-12031-2_6

To simplify the notation, we write $h \to$ id instead of $d_{C^1}(h, \text{id}) \to 0$. Let $\mathcal{F}$ be the collection of C^1 functions $f_h : \mathbb{R} \to \mathbb{R}$ ($h \in \text{Diff}(\Omega_0)$) with the dissipativity condition such that $\overline{d}_{C^1}(f_h, f_0) \leq d_{C^1}(h, \text{id})$, where the metric $\overline{d}_{C^1}$ on $\mathcal{F}$ is given by

$$\overline{d}_{C^1}(f_h, f_{\tilde{h}}) := \min\{d_{C^1}(f_h, f_{\tilde{h}}), 1\} \quad \text{for } h, \tilde{h} \in \text{Diff}(\Omega_0),$$

where id denotes the identity map on Ω_0. For each $h \in \text{Diff}(\Omega_0)$ and $f_h \in \mathcal{F}$, we consider the following perturbation of Eq. (6.1):

$$\begin{cases} \partial_t u - \Delta u = f_h(u) & \text{in } \Omega_h \times (0, \infty), \\ u = 0 & \text{on } \partial\Omega_h \times (0, \infty). \end{cases} \tag{6.2}$$

Then the problem (6.2) is well posed and the Nemytskii operator $F_h : L^2(\Omega_h) \to L^2(\Omega_h)$ of f_h is Lipschitz.

For any $h \in \text{Diff}(\Omega_0)$ C^1-close to id, we consider the equation

$$u_t + A_h u = F_h(u) \tag{6.3}$$

for $u \in L^2(\Omega_h)$, where A_h denotes the operator $-\Delta$ on Ω_h with Dirichlet boundary condition. For simplicity, we write $A_{\text{id}} = A_0$ and $F_{\text{id}} = F_0$. We know that A_h has a sequence of eigenvalues $\{\lambda_i^h\}_{i=1}^\infty$ such that

$$0 < \lambda_1^h \leq \lambda_2^h \leq \cdots \to \infty$$

and a sequence of corresponding eigenfunctions $\{\phi_i^h\}_{i=1}^\infty$, which is an orthonormal basis in $L^2(\Omega_h)$ and orthogonal in $H_0^1(\Omega_h)$. We denote by $S_h(t)$ the semidynamical system induced by Eq. (6.3), defined by

$$S_h(t) : L^2(\Omega_h) \to L^2(\Omega_h), \ S_h(t)(u_0) = u_h(t)$$

for any $t \geq 0$, where $u_h(t)$ is the unique solution of (6.3) with $u_h(0) = u_0$.

For any $h \in \text{Diff}(\Omega_0)$, $f_h \in \mathcal{F}$, and $m \in \mathbb{N}$, let P_m^h be the projection of $L^2(\Omega_h)$ onto $\text{span}\{\phi_1^h, \ldots, \phi_m^h\}$ and Q_m^h be the orthogonal complement of P_m^h. For simplicity, we will write $P_m^{\text{id}} := P_m^0$ and $Q_m^{\text{id}} := Q_m^0$.

Definition 6.1 We say that $\mathcal{M} \subset L^2(\Omega_0)$ is an m-dimensional inertial manifold of the semidynamical system $S_0(t)$ induced by Eq. (6.1) if it is the graph of a Lipschitz map $\Phi : P_m^0 L^2(\Omega_0) \to Q_m^0 L^2(\Omega_0)$ such that

(i) $\mathcal{M}$ is invariant, that is, $S_0(t)\mathcal{M} = \mathcal{M}$ for all $t \in \mathbb{R}$;

(ii) $\mathcal{M}$ attracts all trajectories of $S_0(t)$ exponentially, that is, there are $C > 0$ and $k > 0$ such that for any $u_0 \in L^2(\Omega_0)$, there is $v_0 \in \mathcal{M}$ satisfying

$$\|S_0(t)u_0 - S_0(t)v_0\|_{L^2(\Omega_0)} \le Ce^{-kt}\|u_0 - v_0\|_{L^2(\Omega_0)} \quad \text{for all } t > 0.$$

We are interested in the behavior of the inertial manifolds (which belong to disjoint phase spaces) of Eq. (6.3) with respect to perturbations of the domain Ω_0. To study this, we first need to establish the existence of an inertial manifold of Eq. (6.3) when h is C^1-close enough to id.

Theorem 6.1 *Let the above assumptions on the operator A_h and the nonlinearity F_h hold. In addition, let the following spectral gap condition hold:*

$$\lambda_{m+1}^0 - \lambda_m^0 > 2\sqrt{2}L_0 \quad \text{for some } m \in \mathbb{N}, \tag{6.4}$$

where L_0 is a Lipschitz constant of the nonlinearity F_0 and λ_n^0 is the nth eigenvalue of A_0 for $n \in \mathbb{N}$. Then there exists $\delta > 0$ such that if $d_{C^1}(h, \mathrm{id}) < \delta$, then Eq. (6.3) admits an m-dimensional inertial manifold $\mathcal{M}_h$.

Proof See Theorem 1.1 in [46]. □

Remark 6.1 To the best of our knowledge, the existence of an inertial manifold for (6.1) was first proved by Foias et al. [28] with a non-optimal constant C on the right-hand side of assumption (6.4) (for more details, see Theorem 2.1 in [28]). Moreover, Romanov [66] proved the existence of an inertial manifold of (6.1) under the spectral gap condition (6.4) using the Lyapunov-Perron method in [70]. For a detailed exposition of the classical theory of inertial manifolds, refer to the papers by Zelik [79] and Kostianko and Zelik [42] for a sharp gap condition.

To study how the asymptotic dynamics of evolutionary equation (6.3) changes when we vary the domain Ω_h, our first task is to find a way to compare the inertial manifolds of the equations in different domains. One of the difficulties in this direction is that the phase space $L^2(\Omega_0)$ of the induced semidynamical system changes as we change the domain Ω_0. In fact, the phase spaces $L^2(\Omega_0)$ and $L^2(\Omega_h)$ which contain inertial manifolds $\mathcal{M}_0$ and $\mathcal{M}_h$, respectively, can be disjoint even if Ω_h is a small perturbation of Ω_0.

In this direction, Arrieta and Santamaria [8] estimated the distance of inertial manifolds $\mathcal{M}_s$ of a satisfies evolutionary problem of the form

$$u_t + A_\varepsilon u = F_\varepsilon(u) \tag{6.5}$$

for $\varepsilon \in [0, \varepsilon_0]$ on a Hilbert space X_ε. For this purpose, they first assumed that the operator A_0 satisfies the following spectral gap condition

$$\lambda_{m+1}^0 - \lambda_m^0 \geq 18L_0 \text{ and } \lambda_m^0 \geq 18L_0 \text{ for some } m \in \mathbb{N}$$

to use the Lyapunov-Perron method for the existence of inertial manifolds (see Proposition 2.1 in [8]). They also assumed that the nonlinear terms F_ε have a uniformly bounded support, that is, there exists $R > 0$ such that

$$\text{supp } F_\varepsilon \subset D_R = \{u \in X_\varepsilon : \|u\|_{X_\varepsilon} \leq R\}$$

for $\varepsilon \in [0, \varepsilon_0]$. This assumption implies that the inertial manifold $\mathcal{M}_\varepsilon$ of (6.5) does not leave the ball D_R when ε varies. In fact, we have

$$\mathcal{M}_\varepsilon \cap (X_\varepsilon \setminus D_R) = P_m^\varepsilon(X_\varepsilon) \cap (X_\varepsilon \setminus D_R)$$

for $\varepsilon \in [0, \varepsilon_0]$. Note that the inertial manifold $\mathcal{M}_\varepsilon$ (or $\mathcal{M}_0$) of (6.5) is described by the graph of a Lipschitz map Φ_ε (or Φ_0). Under the above assumptions, they proved that

$$\|\Phi_\varepsilon - E_\varepsilon \Phi_0\|_{L^\infty(\mathbb{R}^m, X_\varepsilon)} \to 0 \quad \text{as } \varepsilon \to 0,$$

where E_ε is an isomorphism from X_0 to X_ε (for more details, see Theorem 2.3 in [8]). Note that the norms $\|\cdot\|_{L^\infty(\mathbb{R}^m, X_\varepsilon)}$ and $\|\cdot\|_{L^\infty(\mathbb{R}^m, X_{\varepsilon'})}$ are not comparable in general if $\varepsilon \neq \varepsilon'$. For any $\varepsilon \in [0, \varepsilon_0]$, we take $h_\varepsilon \in \text{Diff}(\Omega_0)$ satisfying $d_{C^1}(h_\varepsilon, \text{id}) = \varepsilon$. Then the perturbed phase space X_ε in [8] can be considered as the space $L^2(\Omega_{h_\varepsilon})$.

In this chapter, we do not assume that the nonlinear terms F_h ($h \in \text{Diff}(\Omega_0)$) have a uniformly bounded support.

Recently, Lee et al. [49] introduced the Gromov-Hausdorff distance between two dynamical systems on compact metric spaces to analyze how the asymptotic dynamics of the global attractors of (6.1) changes when we vary the domain Ω_0.

To compare the asymptotic behavior of the dynamics on inertial manifolds, we first need to introduce the notion of Gromov-Hausdorff distance between two dynamical systems on noncompact metric spaces. Let (X, d_X) and (Y, d_Y) be two metric spaces. For any $\varepsilon > 0$ and a subset B of X, we recall that a map $i : X \to Y$ is an into ε-isometry on B if $|d_Y(i(x), i(y)) - d_X(x, y)| < \varepsilon$ for all $x, y \in B$. In the case $B = X$, we say that $i : X \to Y$ is an into ε-isometry. An into ε-isometry $i : X \to Y$ is called an ε-isometry if $U_\varepsilon(i(X)) = Y$, where $U_\varepsilon(i(X))$ is the ε-neighborhood of $i(X)$. The Gromov-Hausdorff distance $d_{\text{GH}}(X, Y)$ between X and Y is defined as the infimum of the numbers $\varepsilon > 0$ such that there are ε-isometries $i : X \to Y$ and $j : Y \to X$. Let $\mathcal{X} = \{(X_h, d_{X_h}) : h \in \text{Diff}(\Omega_0)\}$ be a collection of metric spaces. For simplicity, we write X_h instead of (X_h, d_{X_h}). Note that the Gromov-Hausdorff distance $d_{\text{GH}}(X, Y)$ here is equivalent to the Gromov-Hausdorff distance introduced in [14, Definition 7.3.10].

Definition 6.2 We say that $X_h \in \mathcal{X}$ converges to X_k in the Gromov-Hausdorff sense as $h \to k$ if for any $\varepsilon > 0$ and $x_k \in X_k$, there is $\delta > 0$ such that if $d_{C^1}(h, k) < \delta$, then there is $x_h \in X_h$ such that for any $r > 0$,

$$d_{GH}(B(x_h, r), B(x_k, r)) < \varepsilon,$$

where $B(x, r)$ is the closed ball centered at x with radius r.

Note that X_h converges to X_k in the Gromov-Hausdorff sense if

$$d_{GH}(X_h, X_k) \to 0 \quad \text{and} \quad h \to k.$$

However, the converse is not true in general. Let S be a dynamical system on X, namely, $S : X \times \mathbb{R} \to X$. For any subset B of X, we denote by $S|_B$ the restriction of S to $B \times \mathbb{R}$.

Definition 6.3 Let S_1 and S_2 be dynamical systems on metric spaces X and Y, respectively. For any $x \in X$, $y \in Y$, and $r > 0$, the Gromov-Hausdorff distance $D^T_{GH}(S_1|_{B(x,r)}, S_2|_{B(y,r)})$ between $S_1|_{B(x,r)}$ and $S_2|_{B(y,r)}$ with respect to $T > 0$ is defined as the infimum of $\varepsilon > 0$ such that there are maps $i : X \to Y$ and $j : Y \to X$, and reparameterizations $\alpha \in \text{Rep}_{B(x,r)}(\varepsilon)$ and $\beta \in \text{Rep}_{B(y,r)}(\varepsilon)$ with the following properties:

(i) i and j are into ε-isometries on $B(x, r)$ and $B(y, r)$, respectively, satisfying

$$B(y, r) \subset U_\varepsilon(i(B(x, r))) \quad \text{and} \quad B(x, r) \subset U_\varepsilon(j(B(y, r))),$$

(ii) $d_Y(i(S_1(x, \alpha(x, t))), S_2(i(x), t)) < \varepsilon$ for $x \in B(x, r)$ and $t \in [-T, T]$, and $d_X(j(S_2(y, \beta(y, t))), S_1(j(y), t)) < \varepsilon$ for $y \in B(y, r)$ and $t \in [-T, T]$,

where $\text{Rep}_B(\varepsilon)$ is the collection of continuous maps $\alpha : B \times \mathbb{R} \to \mathbb{R}$ such that for each fixed $x \in B$, $\alpha(x, \cdot)$ is a homeomorphism on $\mathbb{R}$ with $\left| \dfrac{\alpha(x, t)}{t} - 1 \right| < \varepsilon$ for $t \neq 0$.

Definition 6.4 Let $\mathcal{DS} = \{(X_h, S_h) : h \in \text{Diff}(\Omega_0)\}$ be a collection of dynamical systems on metric spaces X_h let $x_k \in X_k$. We say that the dynamical system $(X_k, S_k) \in \mathcal{DS}$ is Gromov-Hausdorff stable if for any $\varepsilon > 0$ and $T > 0$, there exists $\delta > 0$ such that if $d_{C^1}(h, k) < \delta$, then there is $x_h \in X_h$ such that for any $r > 0$, $D^T_{GH}(S_h|_{B(x_h,r)}, S_k|_{B(x_k,r)}) < \varepsilon$.

We observe that the Gromov-Hausdorff stability of dynamical systems on the global attractors under perturbations of the domain was first studied in [49].

Throughout this chapter, we assume that the following holds:

$$\lambda_{m+1}^0 - \lambda_m^0 > 2\sqrt{2}L_0 \text{ and } \lambda_m^0 > L_0 \text{ for some } m \in \mathbb{N}. \tag{6.6}$$

Moreover, we assume that m is the smallest number satisfying (6.6) and the inertial manifold $\mathcal{M}_h$ for Eq. (6.3) means the unique m-dimensional inertial manifold for (6.3).

Let us, we state the main results of this section.

Theorem 6.2 *The inertial manifold $\mathcal{M}_h$ of Eq. (6.3) converges to $\mathcal{M}_0$ in the Gromov-Hausdorff sense as $h \to$ id, that is, for any $\varepsilon > 0$ and $u_0 \in \mathcal{M}_0$, there exists $\delta > 0$ such that if $d_{C^1}(h, \text{id}) < \delta$, then there is $u_h \in \mathcal{M}_h$ such that for any $r > 0$, $d_{GH}(B(u_h, r), B(u_0, r)) < \varepsilon$.*

Let $S_h(t)$ be the dynamical system on the inertial manifold $\mathcal{M}_h$ induced by Eq. (6.3).

Theorem 6.3 *The dynamical system $S_0(t)$ is Gromov-Hausdorff stable.*

For the proofs of Theorems 6.2 and 6.3, we first need the continuity of the spectra of A_h with respect to h.

Proposition 6.1 *The spectral data of A_h behave continuously as $h \to$ id. More precisely, for any fixed $\ell \in \mathbb{N}$ and a sequence $\{h_n\}_{n \in \mathbb{N}}$ in $\text{Diff}(\Omega_0)$ with $h_n \to$ id, there exist a subsequence $\{h_k := h_{n_k}\}_{k \in \mathbb{N}}$ of $\{h_n\}_{n \in \mathbb{N}}$ and a collection of eigenfunctions $\{\xi_1^0, \ldots, \xi_\ell^0\}$ of A_0 with corresponding eigenvalues $\{\lambda_1^0, \ldots, \lambda_\ell^0\}$ such that $\lambda_i^{h_k} \to \lambda_i^0$ and $\phi_i^{h_k} \to \xi_i^0$ in $L^2(\mathbb{R}^N)$ for all $1 \le i \le \ell$.*

Proof See Proposition 2.1 in [46]. $\qquad\qquad\qquad\qquad\qquad\qquad\qquad\qquad\qquad\qquad\qquad\qquad\square$

6.2   Proof of Theorem 6.2

For any $h \in \text{Diff}(\Omega_0)$, we let

$$L_h = \sup_{s \in \mathbb{R}} |f_h'(s)| \quad \text{and} \quad L_0 = \sup_{s \in \mathbb{R}} |f_0'(s)|. \tag{6.7}$$

It is clear that L_h and L_0 are Lipschitz constants of the nonlinear terms F_h and F_0, respectively, such that $L_h \to L_0$ as $h \to$ id.

Define a map $j_h : L^2(\Omega_0) \to L^2(\Omega_h)$ by

$$j_h(u) := u \circ h^{-1} \quad \text{for all } u \in L^2(\Omega_0).$$

Then clearly j_h is an isomorphism, and $\|j_h\| \to 1$ as $h \to$ id. Here, $\|j_h\| = \|j_h\|_{L^\infty(L^2(\Omega_0), L^2(\Omega_h))}$. Hence, we may assume that $\|j_h\| < 2$ for all $h \in \text{Diff}(\Omega_0)$.

Now to prove Theorem 6.2, Suppose its conclusion is not true. Then there are $\varepsilon > 0$ and $u_0 \in \mathcal{M}_0$ such that for any $n \in \mathbb{N}$, there is $h_n \in \text{Diff}(\Omega_0)$ with $d_{C^1}(h_n, \text{id}) < 1/n$ such that for any $u_{h_n} \in \mathcal{M}_{h_n}$, there exists $r_n > 0$ such that

$$d_{\text{GH}}(B(u_{h_n}, r_n), B(u_0, r_n)) \geq \varepsilon.$$

For each $n \in \mathbb{N}$, let $\{\lambda_1^{h_n}, \ldots, \lambda_m^{h_n}\}$ and $\{\phi_1^{h_n}, \ldots, \phi_m^{h_n}\}$ be the first m eigenvalues and corresponding m eigenfunctions of A_{h_n}, respectively. By Proposition 6.1, there are m eigenfunctions, denoted by $\{\phi_1^0, \ldots, \phi_m^0\}$, associated to the first m eigenvalues $\{\lambda_1^0, \ldots, \lambda_m^0\}$ of A_0, and a subsequence of $\{h_n\}_{n \in \mathbb{N}}$, still denoted by $\{h_n\}_{n \in \mathbb{N}}$, such that $\lambda_i^{h_n} \to \lambda_i^0$ and $\phi_i^{h_n} \to \phi_i^0$ in $L^2(\mathbb{R}^N)$ as $n \to \infty$ for all $1 \leq i \leq m$. We assume that $|\lambda_i^{h_n} - \lambda_i^0| < 1$ for all $n \in \mathbb{N}$ and $1 \leq i \leq m$. For any $h \in \text{Diff}(\Omega_0)$, define a map $\psi_h : P_m^h L^2(\Omega_h) \to \mathbb{R}^m$ by

$$\psi_h\left(\sum_{i=1}^m a_i \phi_i^h\right) = (a_1, \ldots, a_m), \quad a_i \in \mathbb{R}.$$

For each $n \in \mathbb{N}$, we choose $u_{h_n} \in \mathcal{M}_{h_n}$ such that $\psi_{h_n} P_m^{h_n} u_{h_n} = \psi_0 P_m^0 u_0$.

To complete the proof of Theorem 6.2 we will show that

$$d_{\text{GH}}(B(u_{h_n}, r_n), B(u_0, r_n)) < \varepsilon$$

for all sufficiently large $n \in \mathbb{N}$. For this, we need several lemmas.

Lemma 6.1 *For any fixed $1 \leq i \leq m$, we have*

$$\|j_{h_n} \phi_i^0 - \phi_i^{h_n}\|_{L^2(\Omega_{h_n})} \to 0 \quad as \; n \to \infty.$$

Proof See Lemma 3.1 in [46].   □

Let $\Psi_h : P_m^h L^2(\Omega_h) \to Q_m^h L^2(\Omega_h)$ be the Lipschitz map whose graph is the inertial manifold $\mathcal{M}_h$ in Theorem 6.1. We may assume that Lip $\Psi_h \leq 1$ (see the proof of Theorem 1 in [66]). If we let $\Phi_h = \Psi_h \circ \psi_h^{-1}$, then $\mathcal{M}_h$ can be considered as the graph of Φ_h with Lip $\Phi_h \leq 1$.

With these notations, we have the following lemma.

Lemma 6.2 *For any $p_0 \in P_m^0 L^2(\Omega_0)$ and $p_n \in P_m^{h_n} L^2(\Omega_{h_n})$,*

$$\left|\psi_{h_n} p_n - \psi_0 p_0\right|_{\mathbb{R}^m} \leq \alpha(h_n) \sum_{i=1}^{m} |a_i| + \|j_{h_n} p_0 - p_n\|_{L^2(\Omega_{h_n})},$$

where $\alpha(h_n) = \sup\{\|j_{h_n}\phi_i^0 - \phi_i^{h_n}\|_{L^2(\Omega_{h_n})} : i = 1, \ldots, m\}$ and $p_0 = \sum_{i=1}^{m} a_i \phi_i^0$.

Proof See Lemma 3.2 in [46]. □

By Proposition 6.1, we can take a constant $r > 0$ such that $\lambda_1^0, \lambda_1^{h_n} > r$ for all $n \in \mathbb{N}$. For any $n \in \mathbb{N}$ and $T > 0$, we define

$$\gamma_{h_n}(T) = \sup\{|e^{-\lambda_i^{h_n} t} - e^{-\lambda_i^0 t}| : 1 \leq i \leq m, -T \leq t \leq T\}$$

and

$$\rho(h_n) = \|F_{h_n}(j_{h_n} u) - j_{h_n} F_0(u)\|_{L^\infty(L^2(\Omega_0), L^2(\Omega_{h_n}))}.$$

Then we observe that $\gamma_{h_n}(T) \to 0$ and $\rho(h_n) \to 0$ as $n \to \infty$. For any $p \in \mathbb{R}^m$ and a bounded set $B \subset \mathbb{R}^m$, we define

$$\beta_{h_n}(p) = \|\Phi_{h_n}(p) - j_{h_n}\Phi_0(p)\|_{L^2(\Omega_{h_n})} \text{ and } \beta_{h_n}(B) = \sup\{\beta_{h_n}(p) : p \in B\}.$$

Let $p_0(t)$ and $p_n(t)$ be the solutions of the equations

$$\frac{dp_0}{dt} + A_0 p_0 = P_m^0 F_0(p_0 + \Phi_0(\psi_0 p_0)) \tag{6.8}$$

and

$$\frac{dp_n}{dt} + A_{h_n} p_n = P_m^{h_n} F_{h_n}(p_n + \Phi_{h_n}(\psi_{h_n} p_n)) \tag{6.9}$$

with initial conditions $p_0(0) = \psi_0^{-1} p$ and $p_n(0) = \psi_{h_n}^{-1} p$, respectively, for some $p \in \mathbb{R}^m$. With these notations, we have the following estimates.

Lemma 6.3 *For any $T > 0$ and a bounded subset B of $\mathcal{M}_0$, there exists $C > 0$ such that for any $n \in \mathbb{N}$ and $t \in [-T, 0]$,*

$$\|p_n(t) - j_{h_n} p_0(t)\|_{L^2(\Omega_{h_n})} \leq \left(e^{(\lambda_m^0+1)t} C\gamma_{h_n}(T) + C\alpha(h_n) \right.$$

$$+ \frac{1}{\lambda_m^0 + 1} L_{h_n} \beta_{h_n}(\psi_0 B_{-T}) + \frac{1}{\lambda_m^0 + 1} \rho(h_n)$$

$$+ 2Te^{(\lambda_m^0+1)t} C\gamma_{h_n}(T)$$

$$\left. + \frac{C(2 + L_{h_n})}{\lambda_m^0 + 1} \alpha(h_n) \right) e^{(2L_{h_n} - \lambda_m^0 - 1)t}$$

and for any $n \in \mathbb{N}$ and $t \in [0, T]$,

$$\|p_n(t) - j_{h_n} p_0(t)\|_{L^2(\Omega_{h_n})} \leq \left(e^{rt} C\gamma_{h_n}(T) + C\alpha(h_n) + e^{rt} \frac{L_{h_n}}{r} \beta_{h_n}(\psi_0 B_T) + \right.$$

$$e^{rt} \frac{\rho(h_n)}{r}$$

$$\left. + 2Te^{rt} C\gamma_{h_n}(T) + e^{rt} \frac{C(2 + L_{h_n})}{r} \alpha(h_n) \right) e^{(2L_{h_n} - r)t},$$

$$\|p_n(t) - j_{h_n} p_0(t)\|_{L^2(\Omega_{h_n})} \leq \left(e^{rt} C\gamma_{h_n}(T) + C\alpha(h_n) \right.$$

$$+ e^{rt} \frac{L_{h_n}}{r} \beta_{h_n}(\psi_0 B_T) + e^{rt} \frac{\rho(h_n)}{r} + 2Te^{rt} C\gamma_{h_n}(T)$$

$$\left. + e^{rt} \frac{C(2 + L_{h_n})}{r} \alpha(h_n) \right) e^{(2L_{h_n} - r)t},$$

where $p_0(t)$ and $p_n(t)$ are the solutions of (6.8) and (6.9), respectively, such that $p_0(0) \in P_m^0 B$ and $\psi_0 p_0(0) = \psi_{h_n} p_n(0)$, and $B_{-T} = \{p_0(t) : t \in [-T, 0]\}$, $B_T = \{p_0(t) : t \in [0, T]\}$.

Proof Let $T > 0$ be arbitrary, let B be a bounded subset of $\mathcal{M}_0$, and denote $\hat{B}_{-T} = S_0(B, [-T, 0])$ and $\hat{B}_T = S_0(B, [0, T])$. Let $p_0(t)$ and $p_n(t)$ be the solutions of (6.8) and (6.9), respectively, such that $p_0(0) \in P_m^0 B$ and $\psi_0 p_0(0) = \psi_{h_n} p_n(0)$. By the variation of constants formula for (6.8) and (6.9), we have

$$p_n(t) - j_{h_n} p_0(t) = e^{-A_{h_n} t} p_n(0) - j_{h_n} e^{-A_0 t} p_0(0)$$

$$+ \int_0^t e^{-A_{h_n}(t-s)} (P_m^{h_n} F_{h_n}(p_n + \Phi_{h_n}(\psi_{h_n} p_n)) \qquad (6.10)$$

$$- P_m^{h_n} j_{h_n} F_0(p_0 + \Phi_0(\psi_0 p_0))) ds$$

$$+ \int_0^t (e^{-A_{h_n}(t-s)} P_m^{h_n} j_{h_n} - j_{h_n} e^{-A_0(t-s)} P_m^0)$$

$$F_0(p_0 + \Phi_0(\psi_0 p_0))ds$$

$$:= I + II + III, \quad \text{for all } t \in [-T, T]. \tag{6.11}$$

Since $F_0(\hat{B}_{-T})$ and $F_0(\hat{B}_T)$ are bounded in $L^2(\Omega_0)$, there exists a constant $C > 0$ such that for any $u = \sum_{i=1}^m a_i \phi_i^0$ in $P_m^0 \hat{B}_{-T} \cup P_m^0 \hat{B}_T$ and $v = \sum_{i=1}^m b_i \phi_i^0$ in $P_m^0 F_0(\hat{B}_{-T}) \cup P_m^0 F_0(\hat{B}_T)$, we have $\sum_{i=1}^m |a_i| < C$ and $\sum_{i=1}^m |b_i| < C$.

Step 1. We first estimate I for $t \in [-T, 0]$. We write $p_0(0) = \sum_{i=1}^m a_i \phi_i^0$. Since

$$I = e^{-A_{h_n} t} \sum_{i=1}^m a_i \phi_i^{h_n} - j_{h_n} e^{-A_0 t} \sum_{i=1}^m a_i \phi_i^0$$

$$= \sum_{i=1}^m a_i (e^{-\lambda_i^{h_n} t} - e^{-\lambda_i^0 t}) \phi_i^{h_n} + \sum_{i=1}^m a_i e^{-\lambda_i^0 t} (\phi_i^{h_n} - j_{h_n} \phi_i^0),$$

we have

$$\|I\|_{L^2(\Omega_{h_n})} \leq \left\| \sum_{i=1}^m (e^{-\lambda_i^{h_n} t} - e^{-\lambda_i^0 t}) a_i \phi_i^{h_n} \right\|_{L^2(\Omega_{h_n})}$$

$$+ \sum_{i=1}^m |a_i| e^{-\lambda_i^0 t} \|\phi_i^{h_n} - j_{h_n} \phi_i^0\|_{L^2(\Omega_{h_n})}$$

$$\leq \gamma_{h_n}(T) \sum_{i=1}^m |a_i| + e^{-(\lambda_m^0+1)t} \alpha(h_n) \left(\sum_{i=1}^m |a_i| \right)$$

$$\leq C \gamma_{h_n}(T) + C e^{-(\lambda_m^0+1)t} \alpha(h_n). \tag{6.12}$$

Step 2. We estimate II for $t \in [-T, 0]$. For this, we first consider the following expression:

$$F_{h_n}(p_n + \Phi_{h_n} \psi_{h_n} p_n) - j_{h_n} F_0(p_0 + \Phi_0 \psi_0 p_0)$$

$$= F_{h_n}(p_n + \Phi_{h_n} \psi_{h_n} p_n) - F_{h_n}(j_{h_n} p_0 + \Phi_{h_n} \psi_{h_n} p_n)$$

$$+ F_{h_n}(j_{h_n} p_0 + \Phi_{h_n} \psi_{h_n} p_n) - F_{h_n}(j_{h_n} p_0 + \Phi_{h_n} \psi_0 p_0)$$

$$+ F_{h_n}(j_{h_n} p_0 + \Phi_{h_n} \psi_0 p_0) - F_{h_n}(j_{h_n} p_0 + j_{h_n} \Phi_0 \psi_0 p_0)$$

$$+ F_{h_n}(j_{h_n} p_0 + j_{h_n} \Phi_0 \psi_0 p_0) - j_{h_n} F_0(p_0 + \Phi_0 \psi_0 p_0).$$

By Lemma 6.2, we have

$$\|F_{h_n}(p_n+\Phi_{h_n}\psi_{h_n}p_n) - j_{h_n}F_0(p_0 + \Phi_0\psi_0 p_0)\|_{L^2(\Omega_{h_n})}$$

$$\leq 2L_{h_n}\|p_n(s) - j_{h_n}p_0(s)\|_{L^2(\Omega_{h_n})}$$

$$+ L_{h_n}\beta_{h_n}(\psi_0 B_{-T}) + CL_{h_n}\alpha(h_n) + \rho(h_n).$$

Hence, we get

$$\|III\|_{L^2(\Omega_{h_n})} \leq 2L_{h_n}\int_t^0 e^{-A_{h_n}(t-s)}\|p_n(s) - j_{h_n}p_0(s)\|_{L^2(\Omega_{h_n})}ds$$

$$+ L_{h_n}\int_t^0 e^{-A_{h_n}(t-s)}\beta_{h_n}(\psi_0 B_{-T})ds$$

$$+ CL_{h_n}\int_t^0 e^{-A_{h_n}(t-s)}\alpha(h_n)ds + \int_t^0 e^{-A_{h_n}(t-s)}\rho(h_n)ds$$

$$\leq 2L_{h_n}\int_t^0 e^{-(\lambda_m^0+1)(t-s)}\|p_n(s) - j_{h_n}p_0(s)\|_{L^2(\Omega_{h_n})}ds$$

$$+ L_{h_n}\int_t^0 e^{-(\lambda_m^0+1)(t-s)}\beta_{h_n}(\psi_0 B_{-T})ds$$

$$+ CL_{h_n}\int_t^0 e^{-(\lambda_m^0+1)(t-s)}\alpha(h_n)ds$$

$$+ \int_t^0 e^{-(\lambda_m^0+1)(t-s)}\rho(h_n)ds$$

$$\leq 2L_{h_n}e^{-(\lambda_m^0+1)t}\int_t^0 e^{(\lambda_m^0+1)s}\|p_n(s) - j_{h_n}p_0(s)\|_{L^2(\Omega_{h_n})}ds \qquad (6.13)$$

$$+ L_{h_n}\frac{e^{-(\lambda_m^0+1)t}}{\lambda_m^0 + 1}\beta_{h_n}(\psi_0 B_{-T}) + CL_{h_n}\frac{e^{-(\lambda_m^0+1)t}}{\lambda_m^0 + 1}\alpha(h_n)$$

$$+ \frac{e^{-(\lambda_m^0+1)t}}{\lambda_m^0 + 1}\rho(h_n),$$

where $\int_t^0 e^{(\lambda_m^0+1)s} < \frac{1}{\lambda_m^0+1}$ was used in deriving the last inequality.

Step 3. We estimate III for $t \in [-T, 0]$. To this end, we first consider the following:

$$(j_{h_n}e^{-A_0(t-s)}P_m^0 - e^{-A_{h_n}(t-s)}P_m^{h_n}j_{h_n})(v_0)$$

$$= j_{h_n}\sum_{i=1}^m (e^{-\lambda_i^0(t-s)} - e^{-\lambda_i^{h_n}(t-s)})b_i\phi_i^0 + \sum_{i=1}^m e^{-\lambda_i^{h_n}(t-s)}b_i(j_{h_n}\phi_i^0 - \phi_i^{h_n})$$

$$+ e^{-A_{h_n}(t-s)}(P_m^{h_n} v_n - P_m^{h_n} j_{h_n} v_0)$$

$$:= III_1 + III_2 + III_3,$$

where $v_0 = \sum_{i=1}^{\infty} b_i \phi_i^0 \in F_0(\hat{B}_{-T})$ and $v_n = \sum_{i=1}^{\infty} b_i \phi_i^{h_n} \in L^2(\Omega_{h_n})$. For any $s \in (t, 0]$, we have

$$\|III_1\|_{L^2(\Omega_{h_n})} = \left\| j_{h_n} \sum_{i=1}^{m} (e^{-\lambda_i^0(t-s)} - e^{-\lambda_i^{h_n}(t-s)}) b_i \phi_i^0 \right\|_{L^2(\Omega_{h_n})}$$

$$\leq 2 \left\| \sum_{i=1}^{m} (e^{-\lambda_i^0(t-s)} - e^{-\lambda_i^{h_n}(t-s)}) b_i \phi_i^0 \right\|_{L^2(\Omega_0)}$$

$$\leq 2\gamma_{h_n}(T) \sum_{i=1}^{m} |b_i| \leq 2C\gamma_{h_n}(T),$$

$$\|III_2\|_{L^2(\Omega_{h_n})} = \left\| \sum_{i=1}^{m} e^{-\lambda_i^{h_n}(t-s)} b_i (j_{h_n} \phi_i^0 - \phi_i^{h_n}) \right\|_{L^2(\Omega_{h_n})}$$

$$\leq e^{-(\lambda_m^0+1)(t-s)} \alpha(h_n) \sum_{i=1}^{m} |b_i| \leq C e^{-(\lambda_m^0+1)(t-s)} \alpha(h_n),$$

and

$$\|III_3\|_{L^2(\Omega_{h_n})}^2 = \sum_{i=1}^{m} e^{-2\lambda_i^{h_n}(t-s)} |(P_m^{h_n} v_n - P_m^{h_n} j_{h_n} v_0, \phi_i^{h_n})|^2$$

$$\leq e^{-2(\lambda_m^0+1)(t-s)} \left(\alpha(h_n) \sum_{i=1}^{m} |b_i| \right)^2.$$

It follows that

$$\|III_3\|_{L^2(\Omega_{h_n})} \leq e^{-(\lambda_m^0+1)(t-s)} \alpha(h_n) \sum_{i=1}^{m} |b_i| \leq C e^{-(\lambda_m^0+1)(t-s)} \alpha(h_n).$$

Since $\int_t^0 e^{(\lambda_m^0+1)s} ds < \frac{1}{\lambda_m^0+1}$, we have

$$\|III\|_{L^2(\Omega_{h_n})} \le \int_t^0 2C\gamma_{h_n}(T)ds + \int_t^0 2Ce^{-(\lambda_m^0+1)(t-s)}\alpha(h_n)ds$$

$$\le 2TC\gamma_{h_n}(T) + 2C\alpha(h_n)\frac{e^{-(\lambda_m^0+1)t}}{\lambda_m^0+1}. \tag{6.14}$$

Step 4. We estimate $\|p_n(t) - j_{h_n}p_0(t)\|_{L^2(\Omega_{h_n})}$ for $t \in [-T, 0]$. By putting (6.12), (6.13), and (6.14) together into (6.11), we get

$$\|p_n(t) - j_{h_n}p_0(t)\|_{L^2(\Omega_{h_n})} \le C\gamma_{h_n}(T) + Ce^{-(\lambda_m^0+1)t}\alpha(h_n) \tag{6.15}$$

$$+ 2e^{-(\lambda_m^0+1)t}L_{h_n} \cdot \int_t^0 e^{(\lambda_m^0+1)s}\|p_n(s) - j_{h_n}p_0(s)\|_{L^2(\Omega_{h_n})}ds$$

$$+ \frac{e^{-(\lambda_m^0+1)t}}{\lambda_m^0+1}L_{h_n}\beta_{h_n}(\psi_0 B_{-T}) + \rho(h_n)\frac{e^{-(\lambda_m^0+1)t}}{\lambda_m^0+1}$$

$$+ 2TC\gamma_{h_n}(T) + C(2+L_{h_n})\alpha(h_n)\frac{e^{-(\lambda_m^0+1)t}}{\lambda_m^0+1}. \tag{6.16}$$

Let $g(t) = e^{(\lambda_m^0+1)t}\|p_{h_n}(t) - j_{h_n}p_0(t)\|_{L^2(\Omega_{h_n})}$. Multiply both sides of inequality (6.16) by $e^{(\lambda_m^0+1)t}$ to get

$$g(t) \le e^{(\lambda_m^0+1)t}C\gamma_{h_n}(T) + C\alpha(h_n) + 2L_{h_n}\int_t^0 g(s)ds$$

$$+ \frac{1}{\lambda_m^0+1}L_{h_n}\beta_{h_n}(\psi_0 B_{-T}) + \frac{1}{\lambda_m^0+1}\rho(h_n) + 2Te^{(\lambda_m^0+1)t}C\gamma_{h_n}(T)$$

$$+ \frac{C(2+L_{h_n})}{\lambda_m^0+1}\alpha(h_n).$$

Applying Gronwall's inequality, we derive that

$$g(t) \le \left(e^{(\lambda_m^0+1)t}C\gamma_{h_n}(T) + C\alpha(h_n) + \frac{1}{\lambda_m^0+1}L_{h_n}\beta_{h_n}(\psi_0 B_{-T})\right.$$

$$\left. + \frac{1}{\lambda_m^0+1}\rho(h_n) + 2Te^{(\lambda_m^0+1)t}C\gamma_{h_n}(T) + \frac{C(2+L_{h_n})}{\lambda_m^0+1}\alpha(h_n)\right)e^{2L_{h_n}t}.$$

Consequently, for any $t \in [-T, 0]$, we have

$$\|p_n(t) - j_{h_n}p_0(t)\|_{L^2(\Omega_{h_n})} \le \left(e^{(\lambda_m^0+1)t}C\gamma_{h_n}(T) + C\alpha(h_n)\right.$$

$$+ \frac{1}{\lambda_m^0 + 1} L_{h_n} \beta_{h_n} (\psi_0 B_{-T}) + \frac{1}{\lambda_m^0 + 1} \rho(h_n) + 2T e^{(\lambda_m^0 + 1)t} C \gamma_{h_n}(T)$$

$$+ \frac{C(2 + L_{h_n})}{\lambda_m^0 + 1} \alpha(h_n) \bigg) e^{(-2L_{h_n} - \lambda_m^0 - 1)t}.$$

Step 5. Finally, we estimate $\| p_n(t) - j_{h_n} p_0(t) \|_{L^2(\Omega_{h_n})}$ for $t \in [0, T]$. By the same techniques as in Step 1, we have

$$\| I \|_{L^2(\Omega_{h_n})} \le C \gamma_{h_n}(T) + e^{-rt} C \alpha(h_n).$$

Furthermore, we obtain

$$\| II \|_{L^2(\Omega_{h_n})} \le 2 L_{h_n} \int_0^t e^{-A_{h_n}(t-s)} \| p_n(s) - j_{h_n} p_0(s) \|_{L^2(\Omega_{h_n})} ds$$

$$+ L_{h_n} \beta_{h_n} (\psi_0 B_T) \int_0^t e^{-A_{h_n}(t-s)} ds$$

$$+ C L_{h_n} \alpha(h_n) \int_0^t e^{-A_{h_n}(t-s)} ds + \int_0^t e^{-A_{h_n}(t-s)} \rho(h_n) ds$$

$$\le 2 L_{h_n} e^{-rt} \int_0^t e^{rs} \| p_n(s) - j_{h_n} p_0(s) \|_{L^2(\Omega_{h_n})} ds$$

$$+ L_{h_n} \beta_{h_n} (\psi_0 B_T) e^{-rt} \int_0^t e^{rs} ds$$

$$+ C L_{h_n} \alpha(h_n) e^{-rt} \int_0^t e^{rs} ds + \rho(h_n) e^{-rt} \int_0^t e^{rs} ds$$

$$\le 2 L_{h_n} e^{-rt} \int_0^t e^{rs} \| p_n(s) - j_{h_n} p_0(s) \|_{L^2(\Omega_{h_n})} ds$$

$$+ \frac{L_{h_n}}{r} \beta(h_n) + \frac{C L_{h_n}}{r} \alpha(h_n) + \frac{\rho(h_n)}{r},$$

where the fact that $\int_0^t e^{rs} ds \le e^{rt}/r$ was is used for the last inequality.

To estimate III, we consider separately

$$\| III_1 \|_{L^2(\Omega_{h_n})} \le 2 \gamma_{h_n}(T) \sum_{i=1}^m |b_i| \le 2 C \gamma_{h_n}(T),$$

$$\| III_2 \|_{L^2(\Omega_{h_n})} \le \left\| \sum_{i=1}^m e^{-\lambda_i^h(t-s)} b_i (j_{h_n} \phi_i^0 - \phi_i^{h_n}) \right\|_{L^2(\Omega_{h_n})}$$

$$\le e^{-r(t-s)} \alpha(h_n) \sum_{i=1}^m |b_i| \le e^{-r(t-s)} C \alpha(h_n),$$

and

$$\|III_3\|_{L^2(\Omega_{h_n})} \le e^{-r(t-s)}\alpha(h_n)\sum_{i=1}^{m}|b_i| \le e^{-r(t-s)}C\alpha(h_n).$$

Then we get

$$\|III\|_{L^2(\Omega_{h_n})} \le \int_0^t 2C\gamma_{h_n}(T)ds + \int_0^t 2e^{-r(t-s)}C\alpha(h_n)ds$$

$$\le 2TC\gamma_{h_n}(T) + \frac{2C}{r}\alpha(h_n).$$

Combining all the estimates above we have,

$$\|p_n(t) - j_{h_n}p_0(t)\|_{L^2(\Omega_{h_n})} \le 2e^{-rt}L_{h_n}\int_0^t e^{rs}\|p_n(s) - j_{h_n}p_0(s)\|_{L^2(\Omega_{h_n})}ds$$

$$+ C\gamma_{h_n}(T) + e^{-rt}C\alpha(h_n) + \frac{L_{h_n}}{r}\beta_{h_n}(\psi_0 B_T) + \frac{\rho(h_n)}{r}$$

$$+ 2TC\gamma_{h_n}(T) + \frac{C(2+L_{h_n})}{r}\alpha(h_n). \tag{6.17}$$

Denote $g(t) = e^{rt}\|p_n(t) - j_{h_n}p_0(t)\|_{L^2(\Omega_{h_n})}$. Multiplying both sides of (6.17) by e^{rt} we obtain

$$g(t) \le 2L_{h_n}\int_0^t g(s)ds + e^{rt}C\gamma_{h_n}(T) + C\alpha(h_n) + e^{rt}\frac{L_{h_n}}{r}\beta_{h_n}(\psi_0 B_T)$$

$$+ e^{rt}\frac{\rho(h_n)}{r} + 2Te^{rt}C\gamma_{h_n}(T) + e^{rt}\frac{C(2+L_{h_n})}{r}\alpha(h_n).$$

Next, by Gronwall's inequality,

$$g(t) \le \left(e^{rt}C\gamma_{h_n}(T) + C\alpha(h_n) + e^{rt}\frac{L_{h_n}}{r}\beta_{h_n}(\psi_0 B_T) + e^{rt}\frac{\rho(h_n)}{r}\right.$$

$$\left. + 2Te^{rt}C\gamma_{h_n}(T) + e^{rt}\frac{C(2+L_{h_n})}{r}\alpha(h_n)\right)e^{2L_{h_n}t}.$$

Finally, we deduce that

$$\|p_n(t) - j_{h_n}p_0(t)\|_{L^2(\Omega_{h_n})} \le \left(e^{rt}C\gamma_{h_n}(T) + C\alpha(h_n) + e^{rt}\frac{L_{h_n}}{r}\beta_{h_n}(\psi_0 B_T)\right.$$

$$\left. + e^{rt}\frac{\rho(h_n)}{r} + 2Te^{rt}C\gamma_{h_n}(T) + e^{rt}\frac{C(2+L_{h_n})}{r}\alpha(h_n)\right) \cdot e^{(2L_{h_n}-r)t}.$$

$\square$

In the following lemma, we estimate the linear semigroups on orthogonal complements.

Lemma 6.4 *For any $\varepsilon > 0$, $T > 0$, and a bounded subset B of $L^2(\Omega_0)$, there is $K > 0$ such that for any $u \in B$ and $n \geq K$,*

$$\int_0^T \|e^{-A_{h_n} t} Q_m^{h_n} j_{h_n} u - j_{h_n} e^{-A_0 t} Q_m^0 u\|_{L^2(\Omega_{h_n})} dt < \varepsilon.$$

Proof Since B is bounded, we can choose $\delta > 0$ and $k \in \mathbb{N}$ ($k > m$) such that

$$4\delta \|u\|_{L^2(\Omega_0)} < \varepsilon/2 \quad \text{and} \quad 2e^{-(\lambda_{k+1}^0 - 1)\delta} \|u\|_{L^2(\Omega_0)} < \varepsilon/6(T - \delta)$$

for all $u \in B$. By Proposition 6.1 and Lemma 6.1, we can take a subsequence of $\{h_n\}_{n \in \mathbb{N}}$, still denoted by $\{h_n\}_{n \in \mathbb{N}}$, and the first k eigenfunctions, denoted by $\{\phi_1^0, \ldots, \phi_k^0\}$, with corresponding eigenvalues $\{\lambda_1^0, \ldots, \lambda_k^0\}$, such that

$$\gamma_k(h_n) := \sup\{|e^{-\lambda_i^{h_n} t} - e^{-\lambda_i^0 t}| : 1 \leq i \leq k, 0 \leq t \leq T\} \to 0,$$

and

$$\alpha_k(h_n) := \sup\{\|\phi_i^{h_n} - j_{h_n} \phi_i^0\|_{L^2(\Omega_{h_n})} : 1 \leq i \leq k\} \to 0 \text{ as } n \to \infty.$$

For any $u = \sum_{i=1}^\infty a_i \phi_i^0$ in B and $t \in [0, \delta]$, we have

$$\|e^{-A_{h_n} t} Q_m^{h_n} j_{h_n} u - j_{h_n} e^{-A_0 t} Q_m^0 u\|_{L^2(\Omega_{h_n})}$$

$$\leq \|e^{-A_{h_n} t} Q_m^{h_n} j_{h_n} u\|_{L^2(\Omega_{h_n})} + 2\|e^{-A_0 t} Q_m^0 u\|_{L^2(\Omega_0)}$$

$$\leq 4\|u\|_{L^2(\Omega_0)}.$$

This implies that

$$\int_0^\delta \|e^{-A_{h_n} t} Q_m^{h_n} j_{h_n} u - j_{h_n} e^{-A_0 t} Q_m^0 u\|_{L^2(\Omega_{h_n})} dt < \frac{\varepsilon}{2}. \tag{6.18}$$

For any $t \in [\delta, T]$, we obtain

$$\|e^{-A_{h_n} t} Q_m^{h_n} j_{h_n} u - j_{h_n} e^{-A_0 t} Q_m^0 u\|_{L^2(\Omega_{h_n})}$$

$$\leq \left\| \sum_{i=m+1}^k e^{-\lambda_i^{h_n} t} a_i Q_m^{h_n} j_{h_n} \phi_i^0 - j_{h_n} \sum_{i=m+1}^k e^{-\lambda_i^0 t} a_i \phi_i^0 \right\|_{L^2(\Omega_{h_n})}$$

$$+ \left\| \sum_{i=k+1}^{\infty} e^{-\lambda_i^{h_n} t} a_i Q_m^{h_n} j_{h_n} \phi_i^0 \right\|_{L^2(\Omega_{h_n})} +$$

$$\left\| j_{h_n} \sum_{i=k+1}^{\infty} e^{-\lambda_i^0 t} a_i \phi_i^0 \right\|_{L^2(\Omega_{h_n})}$$

$$:= I + II + III.$$

We first estimate I as follows.

$$I \leq \left\| \sum_{i=m+1}^{k} (e^{-\lambda_i^{h_n} t} - e^{-\lambda_i^0 t}) a_i Q_m^{h_n} j_{h_n} \phi_i^0 \right\|_{L^2(\Omega_{h_n})}$$

$$+ \left\| \sum_{i=m+1}^{k} e^{-\lambda_i^0 t} a_i Q_m^{h_n} j_{h_n} \phi_i^0 - e^{-\lambda_i^0 t} a_i \phi_i^{h_n} \right\|_{L^2(\Omega_{h_n})}$$

$$+ \left\| \sum_{i=m+1}^{k} e^{-\lambda_i^0 t} a_i \phi_i^{h_n} - \sum_{i=m+1}^{k} e^{-\lambda_i^0 t} a_i j_{h_n} \phi_i^0 \right\|_{L^2(\Omega_{h_n})}$$

$$\leq \gamma_k(h_n) \left\| \sum_{i=m+1}^{k} a_i Q_m^{h_n} j_{h_n} \phi_i^0 \right\|_{L^2(\Omega_{h_n})} + 2\alpha_k(h_n) \sum_{i=m+1}^{k} |a_i|.$$

Since $\gamma_k(h_n) \to 0$ and $\alpha_k(h_n) \to 0$ as $n \to \infty$, there exists $K \in \mathbb{N}$ such that if $n \geq K$, then $I < \varepsilon/6(T - \delta)$.

On the other hand, by the choices of δ and k, we have

$$II \leq \|j_{h_n}\| \sum_{i=k+1}^{\infty} e^{-\lambda_i^{h_n} t} |a_i| \leq 2e^{-\lambda_{k+1}^{h_n} \delta} \|u\|_{L^2(\Omega_0)} < \frac{\varepsilon}{6(T - \delta)},$$

and

$$III \leq \|j_{h_n}\| \sum_{i=k+1}^{\infty} e^{-\lambda_i^{h_n} t} |a_i| \leq 2e^{-\lambda_{k+1}^{h_n} \delta} \|u\|_{L^2(\Omega_0)} < \frac{\varepsilon}{6(T - \delta)}.$$

Consequently,

$$\int_{\delta}^{T} \|e^{-A_{h_n} t} Q_m^{h_k} j_{h_k} u - j_{h_k} e^{-A_0 t} Q_m^0 u\|_{L^2(\Omega_{h_k})} dt \leq \int_{\delta}^{T} (I + II + III) dt < \frac{\varepsilon}{2} \quad (6.19)$$

for all $u \in B$. Now (6.18) and (6.19), imply that

$$\int_0^T \|e^{-A_{h_k} t} Q_m^{h_k} j_{h_k} u - j_{h_k} e^{-A_0 t} Q_m^0 u\|_{L^2(\Omega_{h_k})} dt < \varepsilon,$$

which completes the proof.   □

From the proof of Theorem 6.1, we know that $\lambda_{m+1}^h - \lambda_m^h > 2\sqrt{2} L_h$ if $d_{C^1}(h, \mathrm{id})$ is sufficiently small. Then by applying the Lyapunov-Perron method, we see that for $p \in \mathbb{R}^m$ and $n \in \mathbb{N}$,

$$\Phi_{h_n}(p) = \int_{-\infty}^0 e^{A_{h_n} s} Q_m^{h_n} F_{h_n}(p_n(s) + \Phi_{h_n}(\psi_{h_n} p_n(s))) ds \tag{6.20}$$

and

$$\Phi_0(p) = \int_{-\infty}^0 e^{A_0 s} Q_m^0 F_0(p_0(s) + \Phi_0(\psi_0 p_0(s))) ds,$$

where $p_n(t)$ and $p_0(t)$ are the solutions of (6.9) and (6.8) with initial conditions $p_n(0) = \psi_{h_n}^{-1}(p)$ and $p_0(0) = \psi_0^{-1}(p)$, respectively (for more details, see [66]). Since $\overline{d}_{C^1}(f_{h_n}, f_0) \to 0$ as $n \to \infty$, we can take $M_F > 0$ such that, for sufficiently large n,

$$\max\{\|F_0(u_0)\|_{L^2(\Omega_0)}, \|F_{h_n}(u_n)\|_{L^2(\Omega_{h_n})}\} \leq M_F$$

for $u_0 \in L^2(\Omega_0)$ and $u_n \in L^2(\Omega_{h_n})$. Now Theorem 1 in [66], implies that

$$\|\Phi_0(p)\|_{L^2(\Omega_0)} < \frac{M_F}{\lambda_{m+1}^0} \quad \text{and} \quad \|\Phi_{h_n}(p)\|_{L^2(\Omega_{h_n})} < \frac{M_F}{\lambda_{m+1}^{h_n}}.$$

By Proposition 6.1, we can assume that $\lambda_{m+1}^{h_n} \to \lambda_{m+1}^0$ as $n \to \infty$. Then there is $M > 0$ such that

$$\beta_{h_n}(\mathbb{R}^m) < M \tag{6.21}$$

for $n \in \mathbb{N}$. For simplicity, we let F_{h_n} and F_0 denote $F_{h_n} = F_{h_n}(p_n + \Phi_{h_n}(\psi_{h_n} p_n))$ and $F_0 = F_0(p_0 + \Phi_0(\psi_0 p_0))$, respectively.

Lemma 6.5 *For any bounded set $B \subset \mathcal{M}_0$, $\beta_{h_n}(\psi_0 P_m^0 B) \to 0$ as $n \to \infty$.*

Proof Let $\varepsilon > 0$ be arbitrary. Choose $\delta > 0$ such that

$$\eta := \frac{2}{2\sqrt{2}+1-\delta} + \frac{1}{2\sqrt{2}+1-\delta} < 1$$

and denote $\eta_0 = \sum_{i=1}^{\infty} \eta^i$. Take a constant $T > 0$ such that

$$\int_{-\infty}^{-T} \|e^{A_{h_n}s} Q_m^{h_n} F_{h_n} - j_{h_n} e^{A_0 s} Q_m^0 F_0\|_{L^2(\Omega_{h_n})} ds \leq \frac{\varepsilon}{8\eta_0}.$$

For any $k \geq 1$ and a bounded set $B \subset \mathcal{M}_0$, we denote

$$\hat{B}_{-kT} = S_0(B, [-kT, 0]) \quad \text{and} \quad B_{-kT} = P_m^0 S_0(B, [-kT, 0]).$$

Since $F_0(\hat{B}_{-T})$ is bounded in $L^2(\Omega_0)$, there exists a constant $C > 0$ such that for any $u = \sum_{i=1}^{m} a_i \phi_i^0$ in B_{-T} and $v = \sum_{i=1}^{m} b_i \phi_i^0$ in $P_m^0 F_0(\hat{B}_{-T})$, we have $\sum_{i=1}^{m} |a_i| < C$ and $\sum_{i=1}^{m} |b_i| < C$.

Step 1. There is $N_1 > 0$ such that for any $n \geq N_1$,

$$\beta_{h_n}(\psi_0 P_m^0 B) \leq \eta \, \beta_{h_n}(\psi_0 B_{-T}) + \frac{\varepsilon}{4\eta_0}.$$

For any $p \in B$, we have

$$\|\Phi_{h_n}(\psi_0 p) - j_{h_n} \Phi_0(\psi_0 p)\|_{L^2(\Omega_{h_n})}$$

$$\leq \int_{-\infty}^{0} \|e^{A_{h_n}s} Q_m^{h_n} F_{h_n} - j_{h_n} e^{A_0 s} Q_m^0 F_0\|_{L^2(\Omega_{h_n})} ds$$

$$= \int_{-\infty}^{-T} \|e^{A_{h_n}s} Q_m^{h_n} F_{h_n} - j_{h_n} e^{A_0 s} Q_m^0 F_0\|_{L^2(\Omega_{h_n})} ds$$

$$+ \int_{-T}^{0} \|e^{A_{h_n}s} Q_m^{h_n} F_{h_n} - j_{h_n} e^{A_0 s} Q_m^0 F_0\|_{L^2(\Omega_{h_n})} ds$$

$$\leq \frac{\varepsilon}{8\eta_0} + \int_{-T}^{0} \|e^{A_{h_n}s} Q_m^{h_n} (F_{h_n} - j_{h_n} F_0)\|_{L^2(\Omega_{h_n})} ds$$

$$+ \int_{-T}^{0} \|(e^{A_{h_n}s} Q_m^{h_n} j_{h_n} - j_{h_n} e^{A_0 s} Q_m^0) F_0\|_{L^2(\Omega_{h_n})} ds$$

$$:= \frac{\varepsilon}{8\eta_0} + I + II.$$

By Lemma 6.3, we obtain

$$\|e^{A_{h_n}s} Q_m^{h_n} (F_{h_n} - j_{h_n} F_0)\|_{L^2(\Omega_{h_n})} \leq e^{\lambda_{m+1}^{h_n} s} \|(F_{h_n} - j_{h_n} F_0)\|_{L^2(\Omega_{h_n})}$$

$$\leq e^{\lambda_{m+1}^{h_n}s}(2L_{h_n}\|p_n(s) - j_{h_n}p_0(s)\|_{L^2(\Omega_{h_n})} + L_{h_n}\beta_{h_n}(\psi_0 B_{-T})$$

$$+ CL_{h_n}\alpha(h_n) + \rho(h_n))$$

$$\leq \left(2e^{(\lambda_m^0+1)s}L_{h_n}C\gamma_{h_n}(T) + 2L_{h_n}C\alpha(h_n) + \frac{2}{\lambda_m^0+1}L_{h_n}^2\beta_{h_n}(\psi_0 B_{-T})\right.$$

$$+ \frac{2}{\lambda_m^0+1}L_{h_n}\rho(h_n) + 4Te^{(\lambda_m^0+1)s}L_{h_n}C\gamma_{h_n}(T)$$

$$\left.+ \frac{2C(2+L_{h_n})}{\lambda_m^0+1}L_{h_n}\alpha(h_n)\right)e^{(2L_{h_n}+\lambda_{m+1}^{h_n}-\lambda_m^0-1)s}$$

$$+ e^{\lambda_{m+1}^{h_n}s}L_{h_n}\beta_{h_n}(\psi_0 B_{-T}) + e^{\lambda_{m+1}^{h_n}s}CL_{h_n}\alpha(h_n) + e^{\lambda_{m+1}^{h_n}s}\rho(h_n).$$

Consequently,

$$I = \int_{-T}^0 \|e^{A_{h_n}s}Q_m^{h_n}(F_{h_n} - j_{h_n}F_0)\|_{L^2(\Omega_{h_n})}ds$$

$$\leq \frac{2L_{h_n}C\gamma_{h_n}(T)}{2L_{h_n}+\lambda_{m+1}^{h_n}} + \frac{2L_{h_n}C\alpha(h_n)}{2L_{h_n}+\lambda_{m+1}^{h_n}-\lambda_m^0-1}$$

$$+ \frac{2L_{h_n}^2\beta_{h_n}(\psi_0 B_{-T})}{(\lambda_m^0+1)(2L_{h_n}+\lambda_{m+1}^{h_n}-\lambda_m^0-1)} + \frac{2L_{h_n}\rho(h_n)}{2L_{h_n}+\lambda_{m+1}^{h_n}-\lambda_m^0-1}$$

$$+ \frac{4L_{h_n}TC\gamma_{h_n}(T)}{2L_{h_n}+\lambda_{m+1}^{h_n}} + \frac{2C(2+L_{h_n})L_{h_n}\alpha(h_n)}{(\lambda_m^0+1)(2L_{h_n}+\lambda_{m+1}^{h_n}-\lambda_m^0-1)}$$

$$+ \frac{L_{h_n}\beta_{h_n}(\psi_0 B_{-T})}{\lambda_{m+1}^{h_n}} + \frac{CL_{h_n}\alpha(h_n)}{\lambda_{m+1}^{h_n}} + \frac{\rho(h_n)}{\lambda_{m+1}^{h_n}}$$

$$\leq \left(\frac{2L_{h_n}^2}{(\lambda_m^0+1)(2L_{h_n}+\lambda_{m+1}^{h_n}-\lambda_m^0-1)} + \frac{L_{h_n}}{\lambda_{m+1}^{h_n}}\right)\beta_{h_n}(\psi_0 B_{-T})$$

$$+ \frac{2CL_{h_n}\gamma_{h_n}(T)}{2L_{h_n}+\lambda_{m+1}^{h_n}} + \frac{2L_{h_n}C\alpha(h_n)}{2L_{h_n}+\lambda_{m+1}^{h_n}-\lambda_m^0-1}$$

$$+ \frac{2L_{h_n}\rho(h_n)}{2L_{h_n}+\lambda_{m+1}^{h_n}-\lambda_m^0-1} + \frac{4L_{h_n}CT\gamma_{h_n}(T)}{2L_{h_n}+\lambda_{m+1}^{h_n}}$$

$$+ \frac{2C(2+L_{h_n})L_{h_n}\alpha(h_n)}{(\lambda_m^0+1)(2L_{h_n}+\lambda_{m+1}^{h_n}-\lambda_m^0-1)} + \frac{CL_{h_n}\alpha(h_n)}{\lambda_{m+1}^{h_n}} + \frac{\rho(h_n)}{\lambda_{m+1}^{h_n}}$$

$$:= \left(\frac{2L_{h_n}^2}{(\lambda_m^0 + 1)(2L_{h_n} + \lambda_{m+1}^{h_n} - \lambda_m^0 - 1)} + \frac{L_{h_n}}{\lambda_{m+1}^{h_n}} \right) \beta_{h_n}(\psi_0 B_{-T}) + \tilde{I}. \qquad (6.22)$$

By Proposition 6.1, we can take $N_1 > 0$ such that for any $n \geq N_1$,

$$\lambda_m^{h_n} > \lambda_m^0 - \delta \quad \text{and} \quad \lambda_m^{h_n} > L_{h_n} - \delta.$$

Note that $\lambda_m^0 + 1 > L_{h_n}$ and

$$2L_{h_n} + \lambda_{m+1}^{h_n} - \lambda_m^0 - 1 = 2L_{h_n} + (\lambda_{m+1}^{h_n} - \lambda_m^{h_n}) + (\lambda_m^{h_n} - \lambda_m^0) - 1$$
$$> 2L_{h_n} + 2\sqrt{2}L_{h_n} - 1 - \delta.$$

Thus, we have

$$\frac{2L_{h_n}^2}{(\lambda_m^0 + 1)(2L_{h_n} + \lambda_{m+1}^{h_n} - \lambda_m^0 - 1)} \leq \frac{2}{2\sqrt{2} + 1 - \delta}$$

and

$$\frac{L_{h_n}}{\lambda_{m+1}^{h_n}} < \frac{1}{2\sqrt{2} + 1 - \delta}.$$

Since $\gamma_{h_n}(T)$, $\alpha(h_n)$ and $\rho(h_n)$ converge to 0 as $n \to \infty$, we can choose $N_1 > 0$ such that $\tilde{I} < \dfrac{\varepsilon}{16\eta_0}$ for all $n \geq N_1$. Consequently,

$$I < \eta\, \beta_{h_n}(\psi_0 B_{-T}) + \frac{\varepsilon}{16\eta_0}. \qquad (6.23)$$

On the other hand, by Lemma 6.4, we can take $N > 0$ such that for any $u \in L^2(\Omega_0)$ with $\|u\|_{L^2(\Omega_0)} \leq C$ and $n \geq N$,

$$II = \int_{-T}^0 \|(e^{A_{h_n}s} Q_m^{h_n} j_{h_n} - j_{h_n} e^{A_0 s} Q_m^0) F_0\|_{L^2(\Omega_{h_n})} ds < \frac{\varepsilon}{16\eta_0}. \qquad (6.24)$$

By (6.23) and (6.24), we have

$$\|\Phi_{h_n}(p) - j_{h_n}\Phi_0(p)\|_{L^2(\Omega_{h_n})} < \eta\, \beta_{h_n}(\psi_0 B_{-T}) + \frac{\varepsilon}{4\eta_0}.$$

Since p is arbitrary in B, we get

$$\beta_{h_n}(\psi_0 P_m^0 B) < \eta \, \beta_{h_n}(\psi_0 B_{-T}) + \frac{\varepsilon}{4\eta_0}$$

for all $n \geq N_1$. This completes the proof of Step 1.

Step 2. There is $N > 0$ such that $\beta_{h_n}(\psi_0 B) < \varepsilon$ for all $n \geq N$.

Arguing as in Step 1, we conclude that for each $k \in \mathbb{N}$, there is $N_k > N_{k-1}$ such that for any $n \geq N_k$,

$$\beta_{h_n}(\psi_0 B_{-(k-1)T}) < \eta \, \beta_{h_n}(\psi_0 B_{-kT}) + \frac{\varepsilon}{4\eta_0}.$$

Hence, we have

$$\beta_{h_n}(\psi_0 P_m^0 B) < \eta^k \beta_{h_n}(\psi_0 B_{-kT}) + \frac{\varepsilon}{4\eta_0} \sum_{i=0}^{k-1} \eta^i < \eta^k M + \frac{\varepsilon}{4}.$$

Take $k > 0$ such that $\eta^k M < \varepsilon/2$ and $N > N_k$. Then for any $n > N$, we have $\beta_{h_n}(\psi_0 P_m^0 B) < \varepsilon$. This completes the proof. □

For each $n \in \mathbb{N}$, we define a map $\tilde{j}_{h_n} : \mathcal{M}_0 \to \mathcal{M}_{h_n}$ by

$$\tilde{j}_{h_n}(p_0 + \Phi_0(\psi_0 p_0)) = \psi_{h_n}^{-1}\psi_0 p_0 + \Phi_{h_n}(\psi_0 p_0),$$

for $p_0 \in P_m^0 \mathcal{M}_0$. Note that $\tilde{j}_{h_n}$ is continuous. We also define $\hat{j}_{h_n} : B(u_0, r_n) \to B(u_{h_n}, r_n)$ by

$$\hat{j}_{h_n} u = \begin{cases} \tilde{j}_{h_n} u & \text{if } \tilde{j}_{h_n} u \in B(u_{h_n}, r_n), \\ \tilde{j}_{h_n} v & \text{if } \tilde{j}_{h_n} u \notin B(u_{h_n}, r_n), \end{cases}$$

where $v \in B(u_0, r_n)$ is chosen so that

$$\|\tilde{j}_{h_n} u - \tilde{j}_{h_n} v\|_{L^2(\Omega_{h_n})} = dist(\tilde{j}_{h_n} u, \partial B(u_{h_n}, r_n)). \tag{6.25}$$

Similarly, we define $\tilde{i}_{h_n} : \mathcal{M}_{h_n} \to \mathcal{M}_0$ by

$$\tilde{i}_{h_n}(p_n + \Phi_{h_n}(\psi_{h_n} p_n)) = \psi_0^{-1}\psi_{h_n} p_n + \Phi_0(\psi_{h_n} p_n)$$

for $p_n \in P_m^{h_n} \mathcal{M}_{h_n}$ and define $\hat{i}_{h_n} : B(u_{h_n}, r_n) \to B(u_0, r_n)$ by

$$\hat{i}_{h_n} u = \begin{cases} \tilde{i}_{h_n} u & \text{if } \tilde{i}_{h_n} u \in B(u_0, r_n), \\ \tilde{i}_{h_n} v & \text{if } \tilde{i}_{h_n} u \notin B(u_0, r_n), \end{cases}$$

where $v \in B(u_{h_n}, r_n)$ is chosen so that

$$\|\tilde{i}_{h_n} u - \tilde{i}_{h_n} v\|_{L^2(\Omega_0)} = \text{dist}(\tilde{i}_{h_n} u, \partial B(u_0, r_n)). \tag{6.26}$$

Note that $\tilde{i}_{h_n} = \tilde{j}_{h_n}^{-1}$ and $\tilde{i}_{h_n}$ is also continuous.

Lemma 6.6 *Let B be a bounded subset of $\mathcal{M}_0$. Then*

$$\|j_{h_n}(u) - \tilde{j}_{h_n}(u)\|_{L^2(\Omega_{h_n})} \to 0 \quad \text{as } n \to \infty,$$

uniformly for $u \in B$.

Proof Let B be a bounded subset of $\mathcal{M}_0$. For any $u \in B$, there exists $a_i \in \mathbb{R}$ $(1 \leq i \leq m)$ such that

$$u = p_0 + \Phi_0(\psi_0 p_0), \quad \text{with} \quad p_0 = \sum_{i=1}^{m} a_i \phi_i^0 \in P_m^0 \mathcal{M}_0.$$

Since $\|\psi_{h_n}^{-1}\|_{L^\infty(\mathbb{R}^m, P_m^0 L^2(\Omega_{h_n}))} = 1$, implies that Lemma 6.2, we have

$$\begin{aligned}
\|j_{h_n}(u) - \tilde{j}_{h_n}(u)\|_{L^2(\Omega_{h_n})} &= \|j_{h_n}(p_0 + \Phi_0(\psi_0 p_0)) - \tilde{j}_{h_n}(p_0 + \Phi_0(\psi_0 p_0))\|_{L^2(\Omega_{h_n})} \\
&\leq \|j_{h_n} p_0 - \psi_{h_n}^{-1}\psi_0 p_0\|_{L^2(\Omega_{h_n})} + \|j_{h_n}\Phi_0(\psi_0 p_0) \\
&\quad - \Phi_{h_n}(\psi_0 p_0)\|_{L^2(\Omega_{h_n})} \\
&\leq \|\psi_{h_n}^{-1}\|\|\psi_{h_n} j_{h_n} p_0 - \psi_0 p_0|_{\mathbb{R}^m} + \beta_{h_n}(\psi_0 P_m^0 B) \\
&\leq \alpha(h_n) \sum_{i=1}^{m} |a_i| + \beta_{h_n}(\psi_0 P_m^0 B).
\end{aligned}$$

By Lemmas 6.1 and 6.5, we see that $\alpha(h_n)$ and $\beta(h_n)$ converge to 0 as $n \to \infty$. Hence, we derive that $\|j_{h_n}(u) - \tilde{j}_{h_n}(u)\|_{L^2(\Omega_{h_n})} \to 0$ as $n \to \infty$. $\square$

Corollary 6.1 *For $n \in \mathbb{N}$, let B_n be a bounded subset of $\mathcal{M}_{h_n}$. Then*

$$\|i_{h_n}(u) - \tilde{i}_{h_n}(u)\|_{L^2(\Omega_0)} \to 0 \text{ as } n \to \infty$$

uniformly for $u \in B_n$.

Lemma 6.7 *For any $n \in \mathbb{N}$, let $B_n \subset \mathcal{M}_{h_n}$ be a bounded set. Then for any $T > 0$ and $\varepsilon > 0$, there exist $\delta > 0$ and $N_0 \in \mathbb{N}$ such that for all $n \geq N_0$,*

$$\|S_{h_n}(u, t) - S_{h_n}(\tilde{u}, t)\|_{L^2(\Omega_{h_n})} < \varepsilon,$$

uniformly for $u, \tilde{u} \in B_n$ with $\|u - \tilde{u}\|_{L^2(\Omega_{h_n})} < \delta$ and $t \in [-T, T]$.

Proof For a given $n \in \mathbb{N}$, let $B_n \subset \mathcal{M}_{h_n}$ be a bounded set and $u, \tilde{u} \in B_n$. Let $p_n(t)$ and $\tilde{p}_n(t)$ be the solutions of (6.9) with $p_n(0) = P_m^{h_n} u$ and $\tilde{p}_n(0) = P_m^{h_n} \tilde{u}$, respectively. Then

$$S_{h_n}(u, t) = p_n(t) + \Phi_{h_n}(p_n(t)) \text{ and } S_{h_n}(\tilde{u}, t) = \tilde{p}_n(t) + \Phi_{h_n}(\tilde{p}_n(t)). \tag{6.27}$$

Let $p_0(t)$ and $\tilde{p}_0(t)$ be the solutions of Eq. (6.8) with $p_0(0) = \psi_0^{-1} \psi_{h_n} p_n(0)$ and $\tilde{p}_0(0) = \psi_0^{-1} \psi_{h_n} \tilde{p}_n(0)$, respectively. Then we obtain

$$\begin{aligned}
\|p_n(t) - \tilde{p}_n(t)\|_{L^2(\Omega_{h_n})} &\leq \|p_n(t) - j_{h_n} p_0(t)\|_{L^2(\Omega_{h_n})} \\
&\quad + \|j_{h_n} p_0(t) - j_{h_n} \tilde{p}_0(t)\|_{L^2(\Omega_{h_n})} \\
&\quad + \|j_{h_n} \tilde{p}_0(t) - \tilde{p}_n(t)\|_{L^2(\Omega_{h_n})} \\
&:= I + II + III.
\end{aligned} \tag{6.28}$$

By Lemma 6.3, we can choose $N_0 \in \mathbb{N}$ such that if $n \geq N_0$, then

$$I + III \leq \frac{\varepsilon}{4}. \tag{6.29}$$

Moreover, we have

$$II \leq 2\|p_0(t) - \tilde{p}_0(t)\|_{L^2(\Omega_0)}.$$

Since p_0 is Lipschitz on $\psi_0^{-1} \psi_{h_n} P_m^{h_n} B_n \times [-T, T]$, there exists $\delta > 0$ such that if $u, \tilde{u} \in B_n$ with $\|u - \tilde{u}\|_{L^2(\Omega_{h_n})} < \delta$, then

$$II \leq \frac{\varepsilon}{4}. \tag{6.30}$$

Note that

$$\|\Phi_{h_n}(p_n(t)) - \Phi_{h_n}(\tilde{p}_n(t))\|_{L^2(\Omega_{h_n})} \leq \|p_n(t) - \tilde{p}_n(t)\|_{L^2(\Omega_{h_n})}. \tag{6.31}$$

By combining (6.27), (6.28), (6.29), (6.30) and (6.31), we complete the proof of the lemma. $\qquad\square$

Proof (End of Proof of Theorem 6.2)

We first show that there is $N > 0$ such that $\tilde{j}_{h_n}$ is an into ε-isometry on $B(u_0, r_n)$ for all $n \geq N$. Since $B(u_0, r_n)$ is bounded in $L^2(\Omega_0)$, we can take $C_n > 0$ such that $\|u\|_{L^2(\Omega_0)} < C_n$ for all $u \in B(u_0, r_n)$.

By Lemma 6.6, we can take $N > 0$ such that if $n \geq N$, then

$$\left| \|j_{h_n}\| - 1 \right| < \frac{\varepsilon}{18C_n} \text{ and } \|j_{h_n}(u) - \tilde{j}_{h_n}(u)\|_{L^2(\Omega_{h_n})} < \frac{\varepsilon}{9}.$$

For any $u, \tilde{u} \in B(u_0, r_n)$, we let

$$u = p + \Phi_0(\psi_0 p) \text{ and } \tilde{u} = \tilde{p} + \Phi_0(\psi_0 \tilde{p}) \text{ for some } p, \tilde{p} \in P_m^0 \mathcal{M}_0.$$

For any $n \geq N$, we have

$$\|\tilde{j}_{h_n} u - \tilde{j}_{h_n} \tilde{u}\|_{L^2(\Omega_{h_n})} - \|u - \tilde{u}\|_{L^2(\Omega_0)}$$

$$\leq \|\tilde{j}_{h_n}(p + \Phi_0(\psi_0 p)) - j_{h_n}(p + \Phi_0(\psi_0 p))\|_{L^2(\Omega_{h_n})}$$

$$+ \|j_{h_n}(p + \Phi_0(\psi_0 p)) - j_{h_n}(\tilde{p} + \Phi_0(\psi_0 \tilde{p}))\|_{L^2(\Omega_{h_n})}$$

$$+ \|j_{h_n}(\tilde{p} + \Phi_0(\psi_0 \tilde{p})) - \tilde{j}_{h_n}(\tilde{p} + \Phi_0(\psi_0 \tilde{p}))\|_{L^2(\Omega_{h_n})} - \|u - \tilde{u}\|_{L^2(\Omega_0)}$$

$$\leq \frac{2\varepsilon}{9} + (\|j_{h_n}\| - 1)\|u - \tilde{u}\|_{L^2(\Omega_0)} < \frac{\varepsilon}{3}.$$

Similarly, we can show that

$$\|u_0 - \tilde{u}_0\|_{L^2(\Omega_0)} - \|\tilde{j}_{h_n} u_0 - \tilde{j}_{h_n} \tilde{u}_0\|_{L^2(\Omega_{h_n})} < \varepsilon/3.$$

This implies that $\tilde{j}_{h_n}$ is an into $\varepsilon/3$-isometry on $B(u_0, r_n)$. To deduce that $\tilde{i}_{h_n}$ is an into $\varepsilon/3$-isometry on $B(u_{h_n}, r_n)$, we can argue as above and use Corollary 6.1.

Now we show that $\hat{j}_{h_n}$ is an ε-isometry on $B(u_0, r_n)$. Indeed, note that for any $u \in B(u_0, r_n)$,

$$\|\tilde{j}_{h_n} u - u_{h_n}\|_{L^2(\Omega_{h_n})} = \|\tilde{j}_{h_n} u - \tilde{j}_{h_n} u_0\|_{L^2(\Omega_{h_n})} \leq \|u - u_0\|_{L^2(\Omega)} + \frac{\varepsilon}{3} < r_n + \frac{\varepsilon}{3}.$$

This implies that $\tilde{j}_{h_n} B(u_0, r_n) \subset B(u_{h_n}, r_n + \varepsilon/3)$. Hence, for $u, \tilde{u} \in B(u_0, r_n)$, it is enough to consider the following two cases:

Case 1. $\tilde{j}_{h_n} u$ and $\tilde{j}_{h_n} \tilde{u} \notin B(u_{h_n}, r_n)$;
Case 2. $\tilde{j}_{h_n} u \in B(u_{h_n}, r_n)$ and $\tilde{j}_{h_n} \tilde{u} \notin B(u_{h_n}, r_n)$.
For Case 1, by (6.25), we have

$$\|\hat{j}_{h_n} u - \hat{j}_{h_n} \tilde{u}\|_{L^2(\Omega_{h_n})} = \|\tilde{j}_{h_n} v - \tilde{j}_{h_n} \tilde{v}\|_{L^2(\Omega_{h_n})}$$

$$\leq \|\tilde{j}_{h_n} v - \tilde{j}_{h_n} u\|_{L^2(\Omega_{h_n})} + \|\tilde{j}_{h_n} \tilde{v} - \tilde{j}_{h_n} \tilde{u}\|_{L^2(\Omega_{h_n})}$$

$$+ \|\tilde{j}_{h_n} u - \tilde{j}_{h_n} \tilde{u}\|_{L^2(\Omega_{h_n})}$$

$$\leq \varepsilon + \|u - \tilde{u}\|_{L^2(\Omega_0)},$$

where v and $\tilde{v}$ are chosen as in (6.25) corresponding to u and $\tilde{u}$, respectively.

For the Case 2, we see that

$$\|\hat{j}_{h_n} u - \hat{j}_{h_n} \tilde{u}\|_{L^2(\Omega_{h_n})} = \|\tilde{j}_{h_n} u - \tilde{j}_{h_n} \tilde{v}\|_{L^2(\Omega_{h_n})}$$

$$\leq \|\tilde{j}_{h_n} u - \tilde{j}_{h_n} \tilde{u}\|_{L^2(\Omega_{h_n})} + \|\tilde{j}_{h_n} \tilde{v} - \tilde{j}_{h_n} \tilde{u}\|_{L^2(\Omega_{h_n})}$$

$$\leq \frac{2\varepsilon}{3} + \|u - \tilde{u}\|_{L^2(\Omega_0)}.$$

Arguing in the same way, we can show that in any case,

$$\|u - \tilde{u}\|_{L^2(\Omega_0)} \leq \|\hat{j}_{h_n} u - \hat{j}_{h_n} \tilde{u}\|_{L^2(\Omega_{h_n})} + \varepsilon.$$

For any $u_n \in B(u_{h_n}, r_n)$, let $u = \hat{i}_{h_n} u_n$. Then we see that $u \in B(u_0, r_n)$. If $\tilde{j}_{h_n} u \in B(u_{h_n}, r_n)$, then

$$\|u_n - \hat{j}_{h_n} u\|_{L^2(\Omega_{h_n})} = \|u_n - \tilde{j}_{h_n} \hat{i}_{h_n} u_n\|_{L^2(\Omega_{h_n})}$$

$$\leq \|\tilde{i}_{h_n} u_n - \hat{i}_{h_n} u_n\|_{L^2(\Omega_0)} + \frac{\varepsilon}{3}$$

$$\leq \frac{2\varepsilon}{3}.$$

If $\tilde{j}_{h_n} u \notin B(u_{h_n}, r_n)$, then

$$\|u_n - \hat{j}_{h_n} u\|_{L^2(\Omega_{h_n})} = \|u_n - \tilde{j}_{h_n} v\|_{L^2(\Omega_{h_n})}$$

$$\leq \frac{\varepsilon}{3} + \|\tilde{i}_{h_n} u_n - \hat{i}_{h_n} u_n\|_{L^2(\Omega_0)} + \|\tilde{j}_{h_n} \hat{i}_{h_n} u_n - \tilde{j}_{h_n} v\|_{L^2(\Omega_{h_n})}$$

$$\leq \varepsilon.$$

Here, $v \in B(u_0, r_n)$ is chosen as in (6.25) corresponding to $\hat{i}_{h_n} u_n$. This shows that $\hat{j}_{h_n}$ is an ε-isometry on $B(u_0, r_n)$.

Similarly, we can show that $\hat{i}_{h_n}$ is an ε-isometry on $B(u_{h_n}, r_n)$. Consequently, we get

$$d_{GH}(B(u_{h_n}, r_n), B(u_0, r_n)) < \varepsilon$$

for all $n \geq N$. The contradiction completes the proof. $\qquad\square$

6.3 Proof of Theorem 6.3

We are now in a position to prove Theorem 6.3. By the way of contradiction we suppose that the conclusion of the theorem is not true. Then there are $\varepsilon > 0$, $T > 0$, and $u_0 \in \mathcal{M}_0$ such that for any $n \in \mathbb{N}$, there is $h_n \in \text{Diff}(\Omega_0)$ with $d_{C^1}(h_n, id) < 1/n$ such that for any $u_{h_n} \in \mathcal{M}_{h_n}$, there exists $r_n > 0$ such that

$$D_{GH}^T(S_{h_n}|_{B(u_{h_n}, r_n)}, S_0|_{B(u_0, r_n)}) \geq \varepsilon.$$

Let $\{\lambda_1^{h_n}, \ldots, \lambda_m^{h_n}\}$ and $\{\phi_1^{h_n}, \ldots, \phi_m^{h_n}\}$ be the first m eigenvalues and corresponding eigenfunctions of A_{h_n}, respectively. By Proposition 6.1, there are eigenfunctions $\{\phi_1^0, \ldots, \phi_m^0\}$ with respect to the first m eigenvalues $\{\lambda_1^0, \ldots, \lambda_m^0\}$ of A_0 and a subsequence of $\{h_n\}_{n\in\mathbb{N}}$, still denoted by $\{h_n\}_{n\in\mathbb{N}}$, such that

$$\phi_i^{h_n} \to \phi_i^0 \text{ in } L^2(\mathbb{R}^N) \text{ as } n \to \infty \text{ for all } 1 \leq i \leq m.$$

For each $n \in \mathbb{N}$, we choose $u_{h_n} \in \mathcal{M}_{h_n}$ such that $\psi_{h_n} P_m^{h_n} u_{h_n} = \psi_0 P_m^0 u_0$.

Now, let us show that $D_{GH}^T(S_{h_n}|_{B(u_{h_n}, r_n)}, S_0|_{B(u_0, r_n)}) < \varepsilon$ for sufficiently large n. Let $p(t)$ and $p_n(t)$ be the solutions of (6.8) and (6.9), respectively, such that $p(0) \in P_m^0 B(u_0, r_n)$ and $\psi_0 p(0) = \psi_{h_n} p_n(0)$. By Lemma 6.3, there is $N > 0$ such that

$$\|p_n(t)\|_{L^2(\Omega_{h_n})} \leq \|j_{h_n} p(t)\|_{L^2(\Omega_{h_n})} + M$$

for $t \in [-T, T]$ and $n \geq N$, where $M > 0$ is given as in (6.21). It follows that

$$\begin{aligned}
\|\tilde{i}_{h_n}(p_n(t) + \Phi_{h_n}(\psi_{h_n} p_n(t)))\|_{L^2(\Omega_0)} &= \|\psi_0^{-1}\psi_{h_n} p_n(t) + \Phi_0(\psi_{h_n} p_n(t))\|_{L^2(\Omega_0)} \\
&\leq \|p_n(t)\|_{L^2(\Omega_{h_n})} + M \\
&\leq \|j_{h_n} p(t)\|_{L^2(\Omega_{h_n})} + 2M \\
&\leq 2C_n + 2M,
\end{aligned}$$

where $C_n = \sup\{\|S_0(u,t)\|_{L^2(\Omega_0)} : u \in B(u_0, r_n),\ t \in [-T, T]\}$. Then there is $\tilde{u}_0 \in \mathcal{M}_0$ and $\tilde{r}_n > 0$ such that

$$S_0(B(u_0, r_n), [-T, T]) \subset B(\tilde{u}_0, \tilde{r}_n)$$

and

$$\tilde{i}_{h_n}(S_{h_n}(B(u_{h_n}, r_n), [-T, T])) \subset B(\tilde{u}_0, \tilde{r}_n),$$

for all $n \geq N$. For each $n \in \mathbb{N}$, let $\tilde{u}_{h_n} \in \mathcal{M}_{h_n}$ be such that $\psi_{h_n} P_m^{h_n} \tilde{u}_{h_n} = \psi_0 P_m^0 \tilde{u}_0$.

Next we choose $\delta > 0$ and $N_0 \in \mathbb{N}$ corresponding to T, $\varepsilon/4$, and $B(\tilde{u}_{h_n}, \tilde{r}_n + \varepsilon)$ in Lemma 6.7. As in the proof of Theorem 6.2, we can choose $N_1 \in \mathbb{N}$ such that the maps $\hat{j}_{h_n} : B(\tilde{u}_0, \tilde{r}_n) \to B(\tilde{u}_{h_n}, \tilde{r}_n)$ and $\hat{i}_{h_n} : B(\tilde{u}_{h_n}, \tilde{r}_n) \to B(\tilde{u}_0, \tilde{r}_n)$ are $\min\{\delta, \varepsilon/4\}$-isometries for any $n \geq N_1$. Also, we can assume that $\hat{j}_{h_n}$ and $\hat{i}_{h_n}$ are into $\min\{\delta, \varepsilon/4\}$-isometries on $B(\tilde{u}_0, \tilde{r}_n)$ and $B(\tilde{u}_{h_n}, \tilde{r}_n)$, respectively, for all $n \geq N_1$. For a given $T > 0$ and $u \in B(u_0, r_n)$, let $u(t) = S_0(u, t)$ and $u_n(t) = S_{h_n}(\hat{j}_{h_n}(u), t)$ for $t \in [-T, T]$, and denote $B_0^n = \{P_m^0 u(t) : u \in B(u_0, r_n), t \in [-T, T]\}$.

Case 1. Suppose $\tilde{j}_{h_n} S_0(u, t)$ and $\tilde{j}_{h_n} u \in B(\tilde{u}_{h_n}, \tilde{r}_n)$ for $t \in [-T, T]$. Then we have

$$\left\| \hat{j}_{h_n}(S_0(u, t)) - S_{h_n}(\hat{j}_{h_n}(u), t) \right\|_{L^2(\Omega_{h_n})}$$

$$\leq \left\| \tilde{j}_{h_n}(u(t)) - j_{h_n}(u(t)) \right\|_{L^2(\Omega_{h_n})} + \left\| j_{h_n}(u(t)) - u_n(t) \right\|_{L^2(\Omega_{h_n})}$$

$$\leq \left\| \tilde{j}_{h_n}(u(t)) - j_{h_n}(u(t)) \right\|_{L^2(\Omega_{h_n})} + \left\| j_{h_n}(p(t)) - p_n(t) \right\|_{L^2(\Omega_{h_n})}$$

$$+ \left\| j_{h_n} \Phi_0(\psi_0 p(t)) - \Phi_{h_n}(\psi_{h_n} p_n(t)) \right\|_{L^2(\Omega_{h_n})}$$

$$:= I_n + II_n + III_n.$$

By Lemma 6.6, we choose $N_2 > N_1$ such that $I_n < \varepsilon/6$ for any $n \geq N_2$. By Lemma 6.3, we take $N_3 > N_2$ such that

$$II_n = \| j_{h_n}(p(t)) - p_n(t) \|_{L^2(\Omega_{h_n})} < \varepsilon/6$$

for any $n \geq N_3$. Moreover, we have

$$III_n = \| j_{h_n} \Phi_0(\psi_0 p(t)) - \Phi_{h_n}(\psi_{h_n} p_n(t)) \|_{L^2(\Omega_{h_n})}$$

$$\leq \| j_{h_n} \Phi_0(\psi_0 p(t)) - \Phi_{h_n}(\psi_0 p(t)) \|_{L^2(\Omega_{h_n})} + \| \Phi_{h_n}(\psi_0 p(t))$$

$$- \Phi_{h_n}(\psi_{h_n} p_n(t)) \|_{L^2(\Omega_{h_n})}$$

$$\leq \beta_{h_n}(\psi_0 \tilde{B}_0) + \left(\alpha(h_n) \sum_{i=1}^{m} |a_i(t)| + \|j_{h_n} p(t) - p_n(t)\|_{L^2(\Omega_{h_n})} \right)$$

$$= \beta_{h_n}(\psi_0 \tilde{B}_0) + \alpha(h_n) \sum_{i=1}^{m} |a_i(t)| + II_n.$$

Since $\alpha(h_n)$ and $\beta_{h_n}(\psi_0 \tilde{B}_0)$ converge to 0 as $n \to \infty$, by Lemma 6.3, there is an $N_4 > N_3$ such that

$$III_n < \varepsilon/6$$

for all $n \geq N_4$. Consequently,

$$\left\| \hat{j}_{h_n}(S_0(u, t)) - S_{h_n}(\hat{j}_{h_n}(u), t) \right\|_{L^2(\Omega_{h_n})} < \varepsilon \qquad (6.32)$$

for all $n \geq N_4$.

Case 2. Suppose $\tilde{j}_{h_n} S_0(u, t) \in B(\tilde{u}_{h_n}, \tilde{r}_n)$ and $\tilde{j}_{h_n} u \notin B(\tilde{u}_{h_n}, \tilde{r}_n)$ for $t \in [-T, T]$. Then by (6.25) and Lemma 6.7,

$$\left\| \hat{j}_{h_n}(S_0(u, t)) - S_{h_n}(\hat{j}_{h_n}(u), t) \right\|_{L^2(\Omega_{h_n})}$$

$$\leq \left\| \hat{j}_{h_n}(S_0(u, t)) - S_{h_n}(\tilde{j}_{h_n}(u), t) \right\|_{L^2(\Omega_{h_n})}$$

$$+ \left\| S_{h_n}(\tilde{j}_{h_n}(u), t) - S_{h_n}(\hat{j}_{h_n}(u), t) \right\|_{L^2(\Omega_{h_n})}$$

$$\leq \left\| \tilde{j}_{h_n}(u(t)) - j_{h_n}(u(t)) \right\|_{L^2(\Omega_{h_n})} + \left\| j_{h_n}(p(t)) - p_n(t) \right\|_{L^2(\Omega_{h_n})}$$

$$+ \left\| j_{h_n} \Phi_0(\psi_0 p(t)) - \Phi_{h_n}(\psi_{h_n} p_n(t)) \right\|_{L^2(\Omega_{h_n})} + \frac{\varepsilon}{4}.$$

Hence, we can derive (6.32) by arguing as in Case 1.

Case 3. Suppose $\tilde{j}_{h_n} S_0(u, t)$ and $\tilde{j}_{h_n} u \notin B(\tilde{u}_{h_n}, \tilde{r}_n)$ for $t \in [-T, T]$. Then we obtain

$$\left\| \hat{j}_{h_n}(S_0(u, t)) - S_{h_n}(\hat{j}_{h_n}(u), t) \right\|_{L^2(\Omega_{h_n})}$$

$$\leq \left\| \hat{j}_{h_n}(S_0(u, t)) - \tilde{j}_{h_n}(S_0(u, t)) \right\|_{L^2(\Omega_{h_n})} + \left\| \tilde{j}_{h_n}(S_0(u, t)) - S_{h_n}(\tilde{j}_{h_n}(u), t) \right\|_{L^2(\Omega_{h_n})}$$

$$+ \left\| S_{h_n}(\tilde{j}_{h_n}(u), t) - S_{h_n}(\hat{j}_{h_n}(u), t) \right\|_{L^2(\Omega_{h_n})}$$

$$\leq \left\| \tilde{j}_{h_n}(u(t)) - j_{h_n}(u(t)) \right\|_{L^2(\Omega_{h_n})} + \left\| j_{h_n}(p(t)) - p_n(t) \right\|_{L^2(\Omega_{h_n})}$$

$$+ \left\| j_{h_n} \Phi_0(\psi_0 p(t)) - \Phi_{h_n}(\psi_{h_n} p_n(t)) \right\|_{L^2(\Omega_{h_n})} + \frac{\varepsilon}{2}.$$

Hence, we get (6.32) by the same argument as in Case 1.

Similarly, we can prove that

$$\left\| \hat{i}_{h_n}(S_{h_n}(u_n, t)) - S_0(\hat{i}_{h_n}(u_n), t) \right\|_{L^2(\Omega_0)} < \varepsilon$$

for all $n \geq N_4$. This shows that $D_{\mathrm{GH}}^T(S_{h_n}|_{B_{h_n}}, S_0|_{B_0}) < \varepsilon$ for all $n \geq N_4$, and the contradiction we have reached completes the proof. $\square$

Remark 6.2 Suppose the following spectral gap condition for Eq. (6.1) holds:

$$\lambda_{m+1}^0 - \lambda_m^0 > 2L_0 \text{ for some } m \in \mathbb{N}.$$

Then we can prove the continuity and stability of inertial manifolds for reaction-diffusion equations under perturbations of the domain and equation, using a nontrivial generalization of the ODE approach discussed in [67]. For more details, see the first author's forthcoming paper [45].

Stability of Chafee-Infante Equations

7

7.1   Introduction

Chafee and Infante [15] introduced the equation (nowadays called *Chafee-Infante equation*)

$$\begin{cases} u_t - u_{xx} = \lambda f(u) & \text{in } (0, \pi) \times (0, \infty), \\ u(x, t) = 0 & \text{on } \{0, \pi\} \times (0, \infty), \end{cases} \tag{7.1}$$

where $\lambda > 0$ and $f : \mathbb{R} \to \mathbb{R}$ is a C^2 function such that

$$f(0) = 0, \ f' \leq l \quad \text{for some } l > 0, \tag{7.2}$$

Moreover, we assume here that f satisfies the dissipativity condition, namely,

$$\limsup_{|s| \to \infty} \frac{f(s)}{s} < 0.$$

This condition on f easily implies that there exists $C_0 > 0$ such that

$$f(s)s \leq C_0|s| \text{ for all } s \in \mathbb{R}. \tag{7.3}$$

Chafee and Infante analyzed the asymptotic behavior of orbits of the semidynamical system T_0 generated by their equation. Henry [35] proved that if $\lambda \notin \{1^2, 2^2, \ldots\}$, five-one then the time-one map $T_0(1)$ of T_0 is C^2-Morse-Smale, and so it has a finite number of equilibria that are hyperbolic. Throughout the paper, we assume that $\lambda \notin \{1^2, 2^2, \ldots\}$.

© The Author(s), under exclusive license to Springer Nature Switzerland AG 2022
J. Lee, C. Morales, *Gromov-Hausdorff Stability of Dynamical Systems and Applications
to PDEs*, Frontiers in Mathematics, https://doi.org/10.1007/978-3-031-12031-2_7

Very recently, Bortolan et al. [13] dealt with Lipschitz perturbations of Morse-Smale semigroups whose critical elements are hyperbolic. For this, they introduced the notions of $\mathcal{L}$-hyperbolicity, $\mathcal{L}$-transversality, and $\mathcal{L}$-Morse-Smale systems to cases with lack of differentiability.

The objective of this chapter is to prove the geometric stability of Chafee-Infante equations under Lipschitz perturbations of the domain and equation. Moreover, we study the geometric equivalence and the continuity of global attractors. The geometric stability here is a variation of the Gromov-Hausdorff stability in which parametrized families of PDE's are considered. One difficulty one faces when one attacks this problem is that the phase space of the induced semidynamical system changes as we change the domain. In fact, the phase spaces $H_0^1(\Omega)$ and $H_0^1(\Omega_\varepsilon)$, which contain the global attractors $\mathcal{A}$ and $\mathcal{A}_\varepsilon$, respectively, can be disjoint even if Ω_ε is only a small perturbation of $\Omega \subset \mathbb{R}^n$. To overcome this difficulty, many people used the technique adopted by Henry [36], which makes it possible to consider the problem of continuity of the attractors as $\Omega_\varepsilon \to \Omega$ in a fixed phase space $H_0^1(\Omega)$ (see, for example, [11, 60]). More precisely, they followed the general approach, which basically involves "pull-backing" the perturbed problems to the fixed domain Ω and then considering the family of abstract semilinear problems thus generated. Another method is adopted by Arrieta and Carvalho [6], which consider the extended phase space

$$H_\varepsilon^1 := H^1(\Omega_\varepsilon \cap \Omega) \oplus H^1(\Omega_\varepsilon \setminus \overline{\Omega}) \oplus H^1(\Omega \setminus \overline{\Omega}_\varepsilon)$$

to compare the attractors in different phase spaces.

Let us make some remarks about these two methods. First, for the pull-backing technique, the authors considered the pulled-back systems on a fixed domain with induced global attractors and then studied the continuity of the induced global attractors. For the domain extension method in [6], the authors used the norms $\| \cdot \|_{H_\varepsilon^1}$ and $\| \cdot \|_{H_\delta^1}$ on the extended spaces H_ε^1 and H_δ^1 ($\varepsilon, \delta \in [0, \varepsilon_0]$), respectively, which in general cannot be compared if $\varepsilon \neq \delta$. However, we note that it is not easy to obtain some information on the change of dynamics of trajectories inside the global attractors using pull-backing or domain extension methods.

We first perturb the Chafee-Infante Eq. (7.1) as follows. For each $\eta \in [0, 1]$, we consider

$$\begin{cases} u_t - u_{xx} = \lambda f(u) + \eta g(u) & \text{in } \Omega_\eta \times (0, \infty), \\ u(x, t) = 0 & \text{on } \partial\Omega_\eta \times (0, \infty), \end{cases} \tag{7.4}$$

where $g : \mathbb{R} \to \mathbb{R}$ is a globally Lipschitz function with Lipschitz constant L, $\Omega_\eta = (a(\eta), \pi + b(\eta))$, and $a, b : [0, 1] \to \mathbb{R}$ are continuous functions with $a(0) = b(0) = 0$. Note that the nonlinearity $\lambda f + \eta g$ of (7.4) satisfies the dissipativity condition for small η. Then there exists $\eta_0 > 0$ such that for any $\eta \in [0, \eta_0]$, the problem (7.4) is well posed in $H_0^1(\Omega_\eta)$ and has the global attractor $\mathcal{A}_\eta$ (see, for example, Theorems 2.2 and 2.3 in [7]).

For simplicity, we let $X_\eta = L^2(\Omega_\eta)$ with the L^2 norm $\|\cdot\|_{X_\eta}$, and let $X_\eta^{1/2} = H_0^1(\Omega_\eta)$ with the H_0^1 norm $\|\cdot\|_{X_\eta^{1/2}}$. We define a map $i_\eta : X_\eta^{1/2} \to X_0^{1/2}$ by

$$i_\eta u(x) = u\left(x + a(\eta)\frac{1}{\pi}(\pi - x) + b(\eta)\frac{1}{\pi}x\right)$$

and a map $j_\eta : X_0^{1/2} \to X_\eta^{1/2}$ by

$$j_\eta u(x) = u\left(x - \frac{a(\eta)}{\pi + b(\eta) - a(\eta)}(\pi + b(\eta) - x) - \frac{b(\eta)}{\pi + b(\eta) - a(\eta)}(x - a(\eta))\right).$$

By Proposition 3.2 in [7], we see that there exist $\eta_0 > 0$ and $C > 0$ such that for any $\eta \in [0, \eta_0]$ and $\beta \in (\frac{1}{2}, \frac{3}{4})$,

$$\|i_\eta u_\eta\|_{X_\eta^\beta} \leq C,$$

for all $u_\eta \in \mathcal{A}_\eta$. This implies that the set $\bigcup_{\eta \in [0, \eta_0]} i_\eta(\mathcal{A}_\eta)$ is precompact in $X_0^{1/2}$. Moreover, we see that i_η and j_η are continuous, and $\|i_\eta\|_{op} \leq 2$ and $\|j_\eta\|_{op} \leq 2$. Let A_η denote the operator $-\frac{d^2}{dx^2}$ with homogeneous Dirichlet boundary condition on $\partial\Omega_\eta$. Let T_η be the semidynamical system on $X_\eta^{1/2}$ induced by the Lipschitz-perturbed system (7.4). We know that each A_η has a family of eigenvalues $\{\lambda_{k,\eta}\}_{k=1}^\infty$ such that

$$\lambda_{1,\eta} \leq \lambda_{2,\eta} \leq \cdots \to \infty,$$

and a family of corresponding eigenfunctions $\{\phi_{k,\eta}\}_{k=1}^\infty$ contained in $X_\eta^{1/2}$, which is an orthonormal basis in X_η. Let $F_\eta : X_\eta^{1/2} \to X_\eta$ be the Nemytskii operator induced by f and $G_\eta : X_\eta^{1/2} \to X_\eta$ be the Nemytskii operator induced by g. For $\eta \in [0, 1]$ and $\beta \in (\frac{1}{2}, 1)$, we let

$$X_\eta^\beta = \{u \in X_\eta : \|A_\eta^\beta u\|_{X_\eta} < \infty\},$$

where $A_\eta^\beta u = \sum_{k=1}^\infty \lambda_{k,\eta}^\beta (u, \phi_{k,\eta})\phi_{k,\eta}$. Note that $\|A_\eta^\beta \cdot\|_{X_\eta}$ is a norm on X_η^β, which will be denoted by $\|\cdot\|_{X_\eta^\beta}$.

7.2 $\mathcal{L}$-Morse-Smale and Equivalence of Global Attractors

The notions of $\mathcal{L}$-Morse-Smale map and geometric equivalence between two global attractors were first introduced by Bortolan et al. in [13] to investigate the internal dynamics of global attractors cases with lack of differentiability.

Let X be a Banach space, $C(X)$ the set of continuous maps from X into itself, and $\mathcal{L}(X)$ the set of bounded linear operators of X into itself. We say that an equilibrium u^* of $T \in C(X)$ is $\mathcal{L}$-hyperbolic if the map $S : X \to X$ defined by

$$S(u) = T(u + u^*) - T(u^*)$$

has a decomposition of the form $S = L + N$ such that $L \in \mathcal{L}(X)$ is hyperbolic and there is a neighborhood U of 0 such that $N : U \to X$ has a sufficiently small Lipschitz constant. We say that $\xi : \mathbb{Z} \to X$ is a global solution for a map $S \in C(X)$ if $S(\xi(n)) = \xi(n + 1)$ for all $n \in \mathbb{Z}$. Denote by $\mathcal{GS}(S)$ the set of bounded global solutions for $S \in C(X)$. For an $\mathcal{L}$-hyperbolic equilibrium u^* of $S \in C(X)$, let us define

$$W^s(u^*, S) = \{u \in X : S(n)u \to u^* \text{ as } n \to \infty\},$$

$$W^u(u^*, S) = \{u \in X : \exists \xi \in \mathcal{GS}(S) \text{ s.t. } \xi(0) = u \text{ and } \xi(n) \to u^* \text{ as } n \to -\infty\},$$

$$W^s_{\text{loc}}(u^*, S) = \{u - u^* \in U : S(n)u \to u^* \text{ as } n \to \infty\}, \text{ and}$$

$$W^u_{loc}(u^*, S) = \{u - u^* \in U : \exists \xi \in \mathcal{GS}(S) \text{ s.t. } \xi(0) = u \text{ and } \xi(n) \to u^*$$

$$\text{as } n \to -\infty\}.$$

Here, $W^s(u^*, S)$ (resp., $W^u(u^*, S)$) is called the stable (resp., unstable) manifold and $W^s_{\text{loc}}(u^*, S)$ (resp., $W^u_{loc}(u^*, S)$) is called the local stable (resp., local unstable) manifold. Note that if $S \in C(X)$ has two $\mathcal{L}$-hyperbolic equilibria u_1^* and u_2^*, then $W^u(u_1^*, S) \cap W^s_{\text{loc}}(u_2^*, S) \neq \emptyset$ if and only if there exists $\xi \in \mathcal{GS}(S)$ such that

$$\xi(n) \to u_1^* \text{ as } n \to -\infty \quad \text{and} \quad \xi(n) \to u_2^* \text{ as } n \to \infty.$$

In this case, we say that ξ is a connection between u_1^* and u_2^*. Since $T_0(1)$ is Morse-Smale, by the simple application of λ-lemma (see, for example, [58]), we see that for any hyperbolic equilibria u_1^*, u_2^*, and u_3^* of $T_0(1)$,

if $W^u(u_1^*, T_0(1)) \cap W^s_{\text{loc}}(u_2^*, T_0(1))$ and $W^u(u_2^*, T_0(1)) \cap W^s_{\text{loc}}(u_3^*, T_0(1)) \neq \emptyset$,

then $W^u(u_1^*, T_0(1)) \cap W^s_{\text{loc}}(u_3^*, T_0(1)) \neq \emptyset$. $\qquad\qquad$ (7.5)

Let $\mathcal{E}_S = \{u_1^*, \ldots, u_l^*\}$ be the set of $\mathcal{L}$-hyperbolic equilibria of $S \in C(X)$. We say that the map S is dynamically gradient with respect to $\mathcal{E}_S$ if for any $\xi \in \mathcal{GS}$, there exist $u_i^*, u_j^* \in \mathcal{E}_S$ such that

$$\lim_{m \to -\infty} \|\xi(m) - u_i^*\| = 0 \text{ and } \lim_{m \to \infty} \|\xi(m) - u_j^*\| = 0$$

and are there is subset $\{u_{i_1}^*, \ldots, u_{i_p}^*\}$ of $\mathcal{E}_S$ and no elements $\xi_1, \ldots, \xi_p \in \mathcal{GS}$ satisfying

$$\lim_{m \to -\infty} \|\xi_j(m) - u_{i_j}^*\| = 0, \quad \lim_{m \to \infty} \|\xi_j(m) - u_{i_{j+1}}^*\| = 0$$

for $j = 1, \ldots, p$, where $u_{i_{p+1}} = u_{i_1}$.

Let X_1 and X_2 be Banach spaces, and $S_1 \in C(X_1)$ and $S_2 \in C(X_2)$. For the global attractors $\mathcal{A}_i \subset X_i$ of S_i ($i = 1$ or 2), we introduce the notion of geometric equivalence as a generalization of the geometric equivalence introduced in [13].

Definition 7.1 Let $\mathcal{A}_1$ and $\mathcal{A}_2$ be the global attractors of S_1 and S_2, respectively. We say that $\mathcal{A}_1$ is geometrically equivalent to $\mathcal{A}_2$ if S_i is dynamically gradient with respect to its family of $\mathcal{L}$-hyperbolic equilibria $\mathcal{E}_i$ ($i = 1, 2$) and there exists a bijection $\mathcal{B} : \mathcal{E}_1 \to \mathcal{E}_2$ such that

$$W^u(u_i^*, S_1) \cap W_{loc}^s(u_j^*, S_1) \neq \emptyset \text{ iff } W^u(\mathcal{B}(u_i^*), S_2) \cap W_{loc}^s(\mathcal{B}(u_j^*), S_2) \neq \emptyset,$$

where $\mathcal{E}_1 = \{u_1^*, \ldots, u_n^*\}$ and $1 \leq i, j \leq n$.

For any two subsets M, N of X and $x_0 \in M \cap N$, we say that M and N are $\mathcal{L}$-transverse at x_0 if there exist closed subspaces $X_1, X_2 \subset X$ with $X = X_1 \oplus X_2$, a real number $r > 0$, and two Lipschitz functions $\theta : B_r^{X_1}(0) \to X_2$ and $\sigma : B_r^{X_2}(0) \to X_1$ with $\theta(0) = \sigma(0) = 0, \text{Lip}(\theta) < 1, \text{Lip}(\sigma) < 1$, such that

$$\{x_0 + \xi + \theta(\xi) : \xi \in B_r^{X_1}(0)\} \subseteq M \text{ and } \{x_0 + \sigma(\xi) + \xi : \xi \in B_r^{X_2}(0)\} \subseteq N.$$

We denote transversality by $M \pitchfork_{\mathcal{L}, x_0} N$.

Definition 7.2 We say that $S \in C(X)$ is $\mathcal{L}$-Morse-Smale if

(i) S is dynamically gradient with respect to $\mathcal{E}_S = \{u_1^*, \ldots, u_p^*\}$ for $p \in \mathbb{N}$;
(ii) there exists a neighborhood U of $\mathcal{E}_S$ in X such that $S : U \to S(U)$ is bi-Lipschitz;
(iii) if $W^u(u_i^*, S) \cap W_{loc}^s(u_j^*, S) \neq \emptyset$, then there exist $n \in \mathbb{N}$ and $x_0 \in X$ such that

$$T^n W_{loc}^u(u_i^*, S) \pitchfork_{\mathcal{L}, x_0} W_{loc}^s(u_j^*, S) \text{ for } u_i^*, u_j^* \in \mathcal{E}_S \text{ and } 1 \leq i, j \leq p; \text{ and}$$

(iv) (7.5) holds.

In this section, using Theorem 8.9 in [13], we prove that there exists $\eta_0 > 0$ such that for any $\eta \in [0, \eta_0]$, T_η is $\mathcal{L}$-Morse-Smale and the global attractor $\mathcal{A}_\eta$ of T_η is geometrically equivalent to the global attractor $\mathcal{A}_0$ of T_0. For Banach spaces X, Y and a subset U of X, we consider the Lipschitz and sup norms for a map $T : U \to Y$ given by

$$\|T\|_{U,Lip} = \sup_{\substack{x,y\in U \\ x\neq y}} \frac{\|T(x) - T(y)\|}{\|x - y\|} \quad \text{and} \quad \|T\|_{U,\infty} = \sup_{x\in U} \|T(x)\|.$$

Theorem 7.1 (Theorem 8.9 in [13]) *Let $\{S_\eta\}_{\eta\in[0,1]}$ be a family of maps that is collectively asymptotically compact and continuous at $\eta = 0$ in $C(X)$. Suppose that*

(a) *S_η has a global attractor $\mathcal{A}_\eta$ for each $\eta \in [0, 1]$ and $\bigcup_{\eta\in[0,1]} \mathcal{A}_\eta$ is precompact in X;*
(b) *there exists $p \in \mathbb{N}$ such that $\mathcal{E}_\eta(:= \mathcal{E}_{S_\eta}) = \{u^*_{1,\eta}, \ldots, u^*_{p,\eta}\}$ for each $\eta \in [0, 1]$ and*

$$\max_{i=1,\cdots,p} \|u^*_{i,\eta} - u^*_{i,0}\| \to 0 \quad \text{as } \eta \to 0 \quad \text{for } i = 1, \ldots, p;$$

(c) *there exists a neighborhood O of $\bigcup_{\eta\in[0,1]} \mathcal{E}_\eta$ such that*

$$\max\{\|S_\eta - S_0\|_{O,\infty}, \|S_\eta - S_0\|_{O,Lip}\} \to 0 \quad \text{as} \quad \eta \to 0$$

and $S_\eta : O \to S_\eta(O)$ is bi-Lipschitz for each $\eta \in [0, 1]$;
(d) *S_0 is a $\mathcal{L}$-Morse-Smale map and its derivative is uniformly continuous on $O_r(\mathcal{E}_0)$, where $O_r(\mathcal{E}_0)$ is the r-neighborhood of $\mathcal{E}_0$ for $r > 0$.*

Then there exists $\eta_0 > 0$ such that S_η is a $\mathcal{L}$-Morse-Smale map with $\mathcal{A}_\eta$ geometrically equivalent to $\mathcal{A}_0$ for all $\eta \in [0, \eta_0]$.

For any $a \in \mathbb{R}$ and $l > 0$, consider the eigenvalue problem

$$\begin{cases} -u_{xx} = \lambda u & \text{in } (a, a + l), \\ u = 0 & \text{on } \{a, a + l\}. \end{cases}$$

We know that the eigenvalues and corresponding orthonormal eigenfunctions in $L^2(a, a + l)$ of the this problem are given by $\{\left(\frac{n\pi}{l}\right)^2\}_{n\in\mathbb{N}}$ and $\{\sin\frac{n\pi}{l}(\cdot - a)/\|\sin\frac{n\pi}{l}(\cdot - a)\|_{L^2}\}_{n\in\mathbb{N}}$, respectively. Hence, we see that the spectral data of the operator A_0 behave continuously, in the sense that is, for each $k \in \mathbb{N}$, we have

$$\lambda_{k,\eta} \to \lambda_{k,0} \text{ and } i_\eta \phi_{k,\eta} \to \phi_{k,0} \text{ in } X_0^{1/2} \text{ as } \eta \to 0.$$

Furthermore, we can choose $\eta_0 > 0$ and $\delta > 0$ such that $\lambda_{1,\eta} > \delta$ for all $\eta \in [0, \eta_0]$.

We need several lemmas, which are proved as Lemmas 2.2 to 2.9 in [47].

Lemma 7.1 *There exist $M > 0$ (independent of η), $\gamma \in (\frac{1}{2}, 1)$, and a function $\theta :$ $[0, 1] \to \mathbb{R}$ with $\theta(\eta) \to 0$ as $\eta \to 0$ such that for $u_\eta \in L^2(\Omega_\eta)$ and $t > 0$,*

$$\|i_\eta e^{-A_\eta t} u_\eta - e^{-A_0 t} i_\eta u_\eta\|_{X_0^{1/2}} \le M\theta(\eta) t^{-\gamma} e^{-\delta t} \|u_\eta\|_{X_\eta}.$$

Lemma 7.2 *For each $\eta \in [0,1]$, we have the following estimates:*

$$\|T_\eta(t)u_{\eta,0}\|_{X_\eta^{1/2}} \le \sqrt{2\left(\lambda l + \frac{L^2}{\delta}\right) + 1}\, \|u_{\eta,0}\|_{X_\eta} e^{-\frac{\delta}{2}(t-1)} + \sqrt{D_2} \quad \text{for } t \ge 1,$$

$$\|T_\eta(t)u_{\eta,0}\|_{L^\infty(\Omega_\eta)} \le e^{E_1}\|u_{\eta,0}\|_{L^\infty(\Omega_\eta)} + e^{E_1} \quad \text{for } t \in [0,1],$$

where D_2 and E_1 are given in the proof.

Lemma 7.3 *The collection $\{\tilde{T}_\eta(1)\}_{\eta \in [0,1]}$ is collectively asymptotically compact, that is, given sequences $\eta_k \to 0$, $n_k \to \infty$, and $\{u_k\}$ bounded in $X_0^{1/2}$ such that $\{\tilde{T}_{\eta_k}(n_k)u_k\}$ is bounded, $\{\tilde{T}_{\eta_k}(n_k)u_k\}$ has a convergent subsequence.*

Lemma 7.4 *The family $\{\tilde{T}_\eta(1)\}_{\eta \in [0,1]}$ is continuous at $\eta = 0$, that is, for any $N \in \mathbb{N}$ and a compact subset K of $X_0^{1/2}$,*

$$\max_{1 \le n \le N} \sup_{u \in K} \|\tilde{T}_\eta(n)u - T_0(n)u\|_{X_0^{1/2}} \to 0 \quad \text{as } \eta \to 0.$$

Lemma 7.5 *There exists $\eta_0 > 0$ such that for any $T > 0$ and a bounded set U in $X_0^{1/2}$, there is $M_T > 0$ such that*

$$\|T_\eta(t)u_\eta - T_\eta(t)v_\eta\|_{X_\eta^{1/2}} \le M_T \|u_\eta - v_\eta\|_{X_\eta^{1/2}}$$

for all $\eta \in [0, \eta_0]$ and $u_\eta, v_\eta \in j_\eta U$. Here, M_T is independent of η.

Lemma 7.6 *For any bounded set U in $X_0^{1/2}$,*

$$\max\{\|\tilde{T}_\eta(1) - T_0(1)\|_{U,\infty}, \|\tilde{T}_\eta(1) - T_0(1)\|_{U,\text{Lip}}\} \to 0 \quad \text{as } \eta \to 0.$$

Lemma 7.7 *There exist $\eta_0 > 0$ and a neighborhood O of $\overline{\bigcup_{\eta \in [0,\eta_0]} i_\eta \mathcal{A}_\eta}$ such that $\tilde{T}_\eta(1)$ is bi-Lipschitz on O with uniform bi-Lipschitz constants for all $\eta \in [0, \eta_0]$.*

Lemma 7.8 *Let $\mathcal{E}_0 = \{u_{1,0}^*, \ldots, u_{p,0}^*\}$ be the set of equilibria of $T_0(1)$. Then there exists $\eta_0 > 0$ such that for any $\eta \in [0, \eta_0]$, $i_\eta \mathcal{E}_\eta$ consists of p $\mathcal{L}$-hyperbolic equilibria of $\tilde{T}_\eta(1)$, say $i_\eta \mathcal{E}_\eta = \{\tilde{u}_{1,\eta}^*, \ldots, \tilde{u}_{p,\eta}^*\}$, such that*

$$\|\tilde{u}_{i,\eta}^* - u_{i,0}^*\|_{X_0^{1/2}} \to 0 \quad \text{as } \eta \to 0 \text{ for } 1 \le i \le p.$$

With the above lemmas at hand, we are now in a position to prove the main result of this section.

Theorem 7.2 *There is $\eta_0 > 0$ such that if $\eta \in [0, \eta_0]$, then $T_\eta(1)$ is $\mathcal{L}$-Morse-Smale and $\mathcal{A}_\eta$ is geometrically equivalent to $\mathcal{A}_0$.*

Proof Lemmas 7.3 to 7.8, ensure that all assumptions in Theorem 7.1 are satisfied for the collection $\{\tilde{T}_\eta(1)\}_{\eta \in [0,\eta_0]}$, where $\eta_0 > 0$ is a small constant. Hence, $\tilde{T}_\eta(1)$ is $\mathcal{L}$-Morse-Smale and $i_\eta \mathcal{A}_\eta$ is geometrically equivalent to $\mathcal{A}_0$ for all $\eta \in [0, \eta_0]$.

Let $i_\eta \mathcal{E}_\eta = \{\tilde{u}^*_{1,\eta}, \ldots, \tilde{u}^*_{p,\eta}\}$ be the set of $\mathcal{L}$-hyperbolic equilibria of $\tilde{T}_\eta(1)$. For each $1 \leq i \leq p$, we let $j_\eta \tilde{u}^*_{i,\eta} = u^*_{i,\eta}$. Then we see that $\mathcal{E}_\eta := \{u^*_{1,\eta}, \ldots, u^*_{p,\eta}\}$ is the set of all equilibria of $T_\eta(1)$ that are $\mathcal{L}$-hyperbolic. Since $\tilde{T}_\eta(1)$ is dynamically gradient with respect to $i_\eta \mathcal{E}_\eta$, it is clear that $T_\eta(1)$ is dynamically gradient with respect to $\mathcal{E}_\eta$. Let O be a neighborhood of $\overline{\bigcup_{\eta \in [0,\eta_0]} i_\eta \mathcal{A}_\eta}$ in $X_0^{1/2}$ as in Lemma 7.7. Then we see that $j_\eta O$ is a neighborhood of $\mathcal{A}_\eta$ in $X_\eta^{1/2}$ such that $T_\eta(1)$ is bi-Lipschitz on $j_\eta O$. Note that for each $1 \leq i, j \leq p$,

$$W^{\mathrm{u}}(u^*_{i,\eta}, T_\eta(1)) \cap W^{\mathrm{s}}_{\mathrm{loc}}(u^*_{j,\eta}, T_\eta(1)) \neq \emptyset$$

implies

$$W^{\mathrm{u}}(\tilde{u}^*_{i,\eta}, \tilde{T}_\eta(1)) \cap W^{\mathrm{s}}_{\mathrm{loc}}(\tilde{u}^*_{j,\eta}, \tilde{T}_\eta(1)) \neq \emptyset.$$

Since $\tilde{T}_\eta(1)$ is $\mathcal{L}$-Morse-Smale, there exist $n \in \mathbb{N}$ and $u_0 \in X_0^{1/2}$ such that

$$\tilde{T}_\eta(n) W^{\mathrm{u}}_{\mathrm{loc}}(\tilde{u}^*_{i,\eta}, \tilde{T}_\eta(1)) \pitchfork_{\mathcal{L}, u_0} W^{\mathrm{s}}_{\mathrm{loc}}(\tilde{u}^*_{j,\eta}, \tilde{T}_\eta(1)).$$

Consequently,

$$T_\eta(n) W^{\mathrm{u}}_{\mathrm{loc}}(u^*_{i,\eta}, T_\eta(1)) \pitchfork_{\mathcal{L}, j_\eta u_0} W^{\mathrm{s}}_{\mathrm{loc}}(u^*_{j,\eta}, T_\eta(1)).$$

In much the same way one can show that $T_\eta(1)$ satisfies the condition (iv) in Definition 7.2.

On the other hand, since $i_\eta \mathcal{A}_\eta$ is geometrically equivalent to $\mathcal{A}_0$, there is a bijection $\tilde{\mathcal{B}}_\eta : \mathcal{E}_0 \to i_\eta \mathcal{E}_\eta$ such that

$$W^{\mathrm{u}}(u^*_{i,0}, T_0(1)) \cap W^{\mathrm{s}}_{\mathrm{loc}}(u^*_{j,0}, T_0(1)) \neq \emptyset$$

if and only if

$$W^{\mathrm{u}}(\tilde{\mathcal{B}}_\eta(u^*_{i,0}), \tilde{T}_\eta(1)) \cap W^{\mathrm{s}}_{\mathrm{loc}}(\tilde{\mathcal{B}}_\eta(u^*_{j,0}), \tilde{T}_\eta(1)) \neq \emptyset.$$

Let $\mathcal{B}_\eta = j_\eta \circ \tilde{\mathcal{B}}_\eta$. Then $\mathcal{B}_\eta : \mathcal{E}_0 \to \mathcal{E}_\eta$ is a bijection such that

$$W^u(u^*_{i,0}, T_0(1)) \cap W^s_{loc}(u^*_{j,0}, T_0(1)) \neq \emptyset$$

if and only if

$$W^u(\mathcal{B}_\eta(u^*_{i,0}), T_\eta(1)) \cap W^s_{loc}(\mathcal{B}_\eta(u^*_{j,0}), T_\eta(1)) \neq \emptyset.$$

This shows that $\mathcal{A}_\eta$ is geometrically equivalent to $\mathcal{A}_0$, and so completes the proof. $\quad\square$

7.3 Continuity of Global Attractors

We first consider the continuity of local unstable manifolds with respect to the Hausdorff distance.

Lemma 7.9 *For any $\mathcal{L}$-hyperbolic equilibria u^*_η of $T_\eta(1)$ and u^*_0 of $T_0(1)$, if $\|i_\eta u^*_\eta - u^*_0\|_{X^{1/2}_0} \to 0$ as $\eta \to 0$, then*

$$d_H(W^u_{loc}(i_\eta u^*_\eta, \tilde{T}_\eta(1)), W^u_{loc}(u^*_0, T_0(1))) \to 0 \text{ as } \eta \to 0,$$

where $d_H(A, B)$ denotes the Hausdorff distance between A and B for $A, B \subset X^{1/2}_0$.

Proof T the C^2 differentiability of $T_0(1)$ and Lemma 7.6, ensure that all the assumptions of Theorem 5.4 in [13] are satisfied. This completes the proof. $\quad\square$

Theorem 7.3 *The global attractor $\mathcal{A}_\eta$ converges continuously to $\mathcal{A}_0$ in the Gromov-Hausdorff sense, that is, $D_{GH^0}(\mathcal{A}_\eta, \mathcal{A}_0) \to 0$ as $\eta \to 0$.*

Proof Choose $\eta_1 > 0$, $\varepsilon_0 > 0$, and a neighborhood O of $\mathcal{A}_0$ such that for all $\eta \in [0, \eta_1]$,

$$B(\mathcal{A}_0, \varepsilon_0) \subset O \text{ and } i_\eta \mathcal{A}_\eta \subset O,$$

where $B(\mathcal{A}_0, \varepsilon_0)$ denotes the ε_0-neighborhood of $\mathcal{A}_0$ in $X^{1/2}_0$.

Step 1. We first show that for any $0 < \varepsilon < \varepsilon_0$, there is $\eta_2 > 0$ such that $i_\eta \mathcal{A}_\eta \subset B(\mathcal{A}_0, \varepsilon/4)$ for all $\eta \in [0, \eta_2]$.

For any $0 < \varepsilon < \varepsilon_0$, choose $T > 0$ such that $T_0(t)O \subset B(\mathcal{A}_0, \varepsilon/8)$ for all $t \geq T$. By Lemma 7.4, there is $0 < \eta_2 < \eta_1$ such that if $\eta \in [0, \eta_2]$, then

$$\sup_{u \in O} \|\tilde{T}_\eta(T)u - T_0(T)u\|_{X^{1/2}_0} < \frac{\varepsilon}{8}.$$

Since $\|i_\eta\|_{op} \to 1$ as $\eta \to 0$, we have

$$\left| \|i_\eta u_\eta - i_\eta v_\eta\|_{X_0^{1/2}} - \|u_\eta - v_\eta\|_{X_\eta^{1/2}} \right| < \varepsilon/4$$

for $u_\eta, v_\eta \in \mathcal{A}_\eta$. For any $\eta \in [0, \eta_2]$ and $u_\eta \in \mathcal{A}_\eta$, since $\mathcal{A}_\eta$ is invariant, there is $v_\eta \in \mathcal{A}_\eta$ such that $T_\eta(T)v_\eta = u_\eta$. Then we have

$$\mathrm{dist}(i_\eta u_\eta, \mathcal{A}_0) = \mathrm{dist}(i_\eta T_\eta(T)v_\eta, \mathcal{A}_0)$$

$$\leq \|i_\eta T_\eta(T)v_\eta - T_0(T)i_\eta v_\eta\|_{X_0^{1/2}} + \mathrm{dist}(T_0(T)i_\eta v_\eta, \mathcal{A}_0) < \frac{\varepsilon}{4},$$

where dist denotes the Hausdorff semi-distance in $X_0^{1/2}$. Consequently, we have

$$i_\eta \mathcal{A}_\eta \subset B(\mathcal{A}_0, \varepsilon/4) \quad \text{for all } \eta \in [0, \eta_2].$$

Step 2. Next, we show that for any $0 < \varepsilon < \varepsilon_0$, there is $\eta_3 > 0$ such that $j_\eta \mathcal{A}_0 \subset B(\mathcal{A}_\eta, \varepsilon/4)$ for all $\eta \in [0, \eta_3]$.

Suppose this is not the case Then for any $k \in \mathbb{N}$, there are $0 < \eta_k < 1/k$ and $u_k \in \mathcal{A}_0$ such that

$$\mathrm{dist}(j_{\eta_k} u_k, \mathcal{A}_{\eta_k}) \geq \varepsilon/4. \tag{7.6}$$

Since $\mathcal{A}_0$ is compact, we can assume that $u_k \to u_0$ as $k \to \infty$ for some $u_0 \in \mathcal{A}_0$. Take $n \in \mathbb{N}$ such that $T_0(-n)u_0 \in W_{\mathrm{loc}}^u(u_{i,0}^*, T_0(1))$. By Lemma 7.9, there is a sequence of $\tilde{v}_{\eta_k} \in W_{loc}^u(i_{\eta_k} u_{i,\eta_k}^*, \tilde{T}_{\eta_k}(1))$ such that

$$\|\tilde{v}_{\eta_k} - T_0(-n)u_0\|_{X_0^{1/2}} \to 0 \quad \text{as } k \to \infty. \tag{7.7}$$

We have

$$\|\tilde{T}_{\eta_k}(n)\tilde{v}_{\eta_k} - u_0\|_{X_0^{1/2}} \leq \|\tilde{T}_{\eta_k}(n)\tilde{v}_{\eta_k} - T_0(n)\tilde{v}_{\eta_k}\|_{X_0^{1/2}}$$

$$+ \|T_0(n)\tilde{v}_{\eta_k} - T_0(n)T_0(-n)u_0\|_{X_0^{1/2}}.$$

Since $\|T_0(n) - \tilde{T}_{\eta_k}(n)\|_{0,\infty} \to 0$ as $k \to \infty$, by (7.7), we get

$$\|\tilde{T}_{\eta_k}(n)\tilde{v}_{\eta_k} - u_0\|_{X_0^{1/2}} \to 0 \quad \text{as } k \to \infty.$$

Then we obtain

$$\|j_{\eta_k} u_k - T_{\eta_k}(n) j_{\eta_k} \tilde{v}_{\eta_k}\|_{X_{\eta_k}^{1/2}} \leq 2 \|u_k - \tilde{T}_{\eta_k}(n) \tilde{v}_{\eta_k}\|_{X_0^{1/2}}$$

$$\leq 2 \left(\|u_k - u_0\|_{X_0^{1/2}} + \|u_0 - \tilde{T}_{\eta_k}(n) \tilde{v}_{\eta_k}\|_{X_0^{1/2}} \right)$$

$$\to 0$$

as $k \to \infty$. Since $j_{\eta_k} \tilde{v}_{\eta_k} \in \mathcal{A}_{\eta_k}$, we get a contradiction by (7.6).

Step 3. Finally, we show that for any $0 < \varepsilon < \varepsilon_0$, there is $\eta_0 > 0$ such that if $\eta \in [0, \eta_0]$, then there are ε-isometries $\hat{i}_\eta : \mathcal{A}_\eta \to \mathcal{A}_0$ and $\hat{j}_\eta : \mathcal{A}_0 \to \mathcal{A}_\eta$.

For each $\eta \in [0, \eta_2]$, we take a map $\hat{i}_\eta : \mathcal{A}_\eta \to \mathcal{A}_0$ satisfying

$$\hat{i}_\eta u_\eta \in \mathcal{A}_0 \text{ for } u_\eta \in \mathcal{A}_\eta \quad \text{and} \quad \|\hat{i}_\eta u_\eta - i_\eta u_\eta\|_{X_0^{1/2}} < \varepsilon/4.$$

Then $\hat{i}_\eta$ is an ε-isometry. In fact, for any $u_\eta, v_\eta \in \mathcal{A}_\eta$, we have

$$\|\hat{i}_\eta(u_\eta) - \hat{i}_\eta(v_\eta)\|_{X_0^{1/2}} - \|u_\eta - v_\eta\|_{X_\eta^{1/2}} \leq \|\hat{i}_\eta u_\eta - i_\eta u_\eta\|_{X_0^{1/2}}$$

$$+ \|i_\eta v_\eta - \hat{i}_\eta v_\eta\|_{X_0^{1/2}} + \|i_\eta u_\eta$$

$$- i_\eta v_\eta\|_{X_0^{1/2}} - \|u_\eta - v_\eta\|_{X_\eta^{1/2}}$$

$$< \varepsilon.$$

Similarly, we can show that

$$\|u_\eta - v_\eta\|_{X_\eta^{1/2}} - \|\hat{i}_\eta u_\eta - \hat{i}_\eta v_\eta\|_{X_0^{1/2}} < \varepsilon.$$

By Step 2, for $u_0 \in \mathcal{A}_0$, we can choose $u_\eta \in \mathcal{A}_\eta$ such that $\|i_\eta u_\eta - u_0\|_{X_0^{1/2}} < \varepsilon/2$ for all $\eta \in [0, \eta_2]$. Since

$$\|\hat{i}_\eta u_\eta - u_0\|_{X_0^{1/2}} \leq \|\hat{i}_\eta u_\eta - i_\eta u_\eta\|_{X_0^{1/2}} + \|i_\eta u_\eta - u_0\|_{X_0^{1/2}} < \varepsilon,$$

we get $\mathcal{A}_0 \subset B(\hat{i}_\eta(\mathcal{A}_\eta), \varepsilon)$. Hence, $\hat{i}_\eta$ is an ε-isometry.

Take $0 < \eta_4 < \eta_3$ such that if $\eta \in [0, \eta_4]$. Then

$$\left| \|j_\eta u_0 - j_\eta v_0\|_{X_\eta^{1/2}} - \|u_0 - v_0\|_{X_0^{1/2}} \right| < \varepsilon/4$$

for all $u_0, v_0 \in \mathcal{A}_0$. For each $\eta \in [0, \eta_4]$, take a map $\hat{j}_\eta : \mathcal{A}_0 \to \mathcal{A}_\eta$ satisfying

$$\hat{j}_\eta u_0 \in \mathcal{A}_\eta \text{ as } u_0 \in \mathcal{A}_0 \quad \text{and} \quad \|\hat{j}_\eta u_0 - j_\eta u_0\|_{X_\eta^{1/2}} < \varepsilon/4.$$

Then $\hat{j}_\eta$ is an ε-isometry. In fact, for any $u_0, v_0 \in \mathcal{A}_0$, we have

$$\|\hat{j}_\eta u_0 - \hat{j}_\eta v_0\|_{X_\eta^{1/2}} - \|u_0 - v_0\|_{X_0^{1/2}} \leq \|\hat{j}_\eta u_0 - j_\eta u_0\|_{X_\eta^{1/2}}$$

$$+ \|j_\eta v_0 - \hat{j}_\eta v_0\|_{X_\eta^{1/2}} + \|j_\eta u_0 - j_\eta v_0\|_{X_\eta^{1/2}} - \|u_0 - v_0\|_{X_0^{1/2}}$$

$$< \varepsilon.$$

Similarly, we can show that

$$\|u_0 - v_0\|_{X_0^{1/2}} - \|\hat{j}_\eta u_0 - \hat{j}_\eta v_0\|_{X_\eta^{1/2}} < \varepsilon.$$

By Step 1, for $u_\eta \in \mathcal{A}_\eta$, we can choose $u_0 \in \mathcal{A}_0$ such that $\|j_\eta u_0 - u_\eta\|_{X_\eta^{1/2}} < \varepsilon/2$ for any $\eta \in [0, \eta_4]$. Since

$$\|\hat{j}_\eta u_0 - u_\eta\|_{X_\eta^{1/2}} \leq \|\hat{j}_\eta u_0 - j_\eta u_0\|_{X_\eta^{1/2}} + \|j_\eta u_0 - u_\eta\|_{X_\eta^{1/2}} < \varepsilon,$$

we get $\mathcal{A}_\eta \subset B(\hat{j}_\eta(\mathcal{A}_0), \varepsilon)$. Hence, $\hat{j}_\eta$ is an ε-isometry.

Let $\eta_0 = \min\{\eta_2, \eta_4\}$. Then we see that for any $\eta \in [0, \eta_0]$, $\hat{i}_\eta : \mathcal{A}_\eta \to \mathcal{A}_0$ and $\hat{j}_\eta : \mathcal{A}_0 \to \mathcal{A}_\eta$ are ε-isometries. Consequently, $D_{\mathrm{GH}^0}(\mathcal{A}_\eta, \mathcal{A}_0) < \varepsilon$ for all $\eta \in [0, \eta_0]$. $\qquad\square$

7.4 Geometric Stability

It is well known that every Morse-Smale dynamical system on a compact smooth manifold M is structurally stable under C^r-perturbations ($r \geq 1$), that is, there is a homeomorphism on M which sends the orbits in the perturbed system to the orbits in the original system (see, for example, [32, 33]).

In what follows we study the stability of T_0 on the global attractor $\mathcal{A}_0$ under Lipschitz perturbations of the domain and equation in (7.1). To this end, we introduce the notion of geometric stability of (7.1).

Definition 7.3 We say that the system (7.1) is geometrically stable if for any $\varepsilon > 0$, there exists $\eta_0 > 0$ such that for any $\eta \in [0, \eta_0]$, there exist an ε-isometry $\tilde{i}_\eta : \mathcal{A}_\eta \to \mathcal{A}_0$ and $\alpha \in \mathrm{Rep}_{\mathcal{A}_\eta}$ such that for any $u \in \mathcal{A}_\eta$ and $t \in \mathbb{R}$,

$$\tilde{i}_\eta T_\eta(\alpha(u, t))u = T_0(t)\tilde{i}_\eta u.$$

We can now state our main result of in this section.

Theorem 7.4 *The dynamical system T_0 on the global attractor $\mathcal{A}_0$ is geometrically stable. More precisely, for any $\varepsilon > 0$, there is $\eta_0 > 0$ such that if $\eta \in [0, \eta_0]$, then there is an ε-isometry $\tilde{i}_\eta : \mathcal{A}_\eta \to \mathcal{A}_0$ such that for any $u_\eta \in \mathcal{A}_\eta$ and $t \in \mathbb{R}$,*

$$\tilde{i}_\eta T_\eta(t) u_\eta = T_0(t) \tilde{i}_\eta u_\eta.$$

Moreover, $\tilde{i}_\eta|_{\mathcal{E}_\eta} : \mathcal{E}_\eta \to \mathcal{E}_0$ is a bijection such that

$$u_\eta \in W^u(u^*_{i,\eta}, T_\eta(1)) \cap W^s_{\mathrm{loc}}(u^*_{j,\eta}, T_\eta(1)) \text{ if and only if}$$

$$\tilde{i}_\eta u_\eta \in W^u(\tilde{i}_\eta u^*_{i,\eta}, T_0(1)) \cap W^s_{\mathrm{loc}}(\tilde{i}_\eta u^*_{j,\eta}, T_0(1)),$$

where $1 \le i, j \le p$. Here, $\mathcal{E}_\eta := \{u^*_{1,\eta}, \ldots, u^*_{p,\eta}\}$ is the set of all $\mathcal{L}$-hyperbolic equilibria of $T_\eta(1)$.

Remark 7.1 Theorem 7.2, implies that if there is a connection in $X_\eta^{1/2}$, then there is also a connection in $X_0^{1/2}$, and vice versa. However, Theorem 7.4 gives us more information. In fact, we see that $\tilde{i}_\eta$ maps a connection in $X_\eta^{1/2}$ into a connection in $X_0^{1/2}$.

We observe that the system (7.4) induces a dynamical system T_η on its global attractor $\mathcal{A}_\eta$ for sufficient small η. In fact, for a sufficiently small constant $\eta_0 > 0$, let $\tilde{F}_\eta : X_\eta^{1/2} \to X_\eta$ be a truncation of F_η such that

$$\tilde{F}_\eta(u) = \begin{cases} F_\eta(u) & \text{for } u \in B(\mathcal{A}_\eta, \varepsilon_\eta), \\ 0 & \text{for } u \notin B(\mathcal{A}_\eta, 2\varepsilon_\eta), \end{cases}$$

and $\tilde{F}_\eta$ is Lipschitz, where $\eta \in [0, \eta_0]$ and $\varepsilon_\eta > 0$. Since G_η is Lipschitz, the semidynamical system S_η induced by the equation

$$\frac{du}{dt} + A_\eta u = \tilde{F}_\eta(u) + G_\eta(u)$$

has an inertial manifold $\mathcal{M}_\eta$ that is invariant under S_η, that is, $S_\eta(t)\mathcal{M}_\eta = \mathcal{M}_\eta$ for all $t \in \mathbb{R}$ (see, for example, [28, 53]). Since the global attractor $\mathcal{A}_\eta$ of T_η is contained in $\mathcal{M}_\eta$ and $S_\eta|_{\mathcal{A}_\eta} = T_\eta|_{\mathcal{A}_\eta}$, we see that T_η is a dynamical system on $\mathcal{A}_\eta$ for each $\eta \in [0, \eta_0]$.

To prove Theorem 7.4, we first study the stability in the Gromov-Hausdorff sense of the semidynamical system T_0 induced by (7.1). For this, let us recall the concept of Gromov-Hausdorff distance between two dynamical systems on compact metric spaces, introduced by Lee et al. in [49]. Let $\mathcal{CDS}$ be the collection of all dynamical systems on compact metric spaces up to isometry. For two systems $(X, \phi), (Y, \psi) \in \mathcal{CDS}$ and $T > 0$, the Gromov-Hausdorff distance $D_{\mathrm{GH}^0}^T(\phi, \psi)$ is defined as the infimum of ε for which that there exist two ε-isometries $i : X \to Y$ and $j : Y \to X$, and reparameterizations $\alpha \in \mathrm{Rep}_X$ and $\beta \in \mathrm{Rep}_Y$ such that for any $x \in X$, $y \in Y$ and $t \in [-T, T]$,

$$d(i(\phi(x, \alpha(x, t))), \psi(i(x), t)) < \varepsilon \text{ and } d(j(\psi(y, \beta(y, t))), \phi(j(y), t)) < \varepsilon;$$

recall that Rep_X denotes the set of all continuous functions $\alpha : X \times \mathbb{R} \to \mathbb{R}$ such that for each fixed $x \in X$, $\alpha(x, \cdot) : \mathbb{R} \to \mathbb{R}$ is an increasing homeomorphism with $\alpha(x, 0) = 0$.

Using the distance $D_{\mathrm{GH}^0}^T$, we introduce the Gromov-Hausdorff stability of the system (7.1) as follows.

Definition 7.4 We say that the system (7.1) is Gromov-Hausdorff stable if for any $T > 0$ and $\varepsilon > 0$, there exists $\eta_0 > 0$ such that for any $\eta \in [0, \eta_0]$, there exist ε-isometries $\hat{i}_\eta : \mathcal{A}_\eta \to \mathcal{A}_0$, $\hat{j}_\eta : \mathcal{A}_0 \to \mathcal{A}_\eta$, and reparameterizations $\alpha \in \mathrm{Rep}_{\mathcal{A}_\eta}$, $\beta \in \mathrm{Rep}_{\mathcal{A}_0}$ such that

$$\|\hat{i}_\eta T_\eta(\alpha(u, t))u - T_0(t)\hat{i}_\eta u\|_{X_0^{1/2}} < \varepsilon \text{ and } \|\hat{j}_\eta T_0(\beta(v, t))v - T_\eta(t)\hat{j}_\eta v\|_{X_\eta^{1/2}} < \varepsilon,$$

for all $u \in \mathcal{A}_\eta$, $v \in \mathcal{A}_0$, and $t \in [-T, T]$.

To prove the Gromov-Hausdorff stability of the system (7.1), we need the following result.

Lemma 7.10 *For any $T > 0$ and $\varepsilon > 0$, there are $\eta_0 > 0$ and $\delta > 0$ such that for any $\eta \in [0, \eta_0]$, if $\|u_\eta - v_\eta\|_{X_\eta^{1/2}} < \delta$ for $u_\eta, v_\eta \in \mathcal{A}_\eta$ and $|t - s| < \delta$ for $s, t \in [-T, T]$, then $\|T_\eta(t)u_\eta - T_\eta(s)v_\eta\|_{X_\eta^{1/2}} < \varepsilon$.*

Proof See Lemma 4.2 in [47].   □

Theorem 7.5 *The dynamical system T_0 on the global attractor $\mathcal{A}_0$ is Gromov-Hausdorff stable.*

Proof Step 1. We first show that for any $T > 0$ and $\varepsilon > 0$, there is $\eta_3 > 0$ such that for any $\eta \in [0, \eta_3]$, there are $\varepsilon/3$-isometries

$$\hat{i}_\eta : \mathcal{A}_\eta \to \mathcal{A}_0 \text{ and } \hat{j}_\eta : \mathcal{A}_0 \to \mathcal{A}_\eta$$

such that for any $u_0 \in \mathcal{A}_0$, $u_\eta \in \mathcal{A}_\eta$, and $t \in [0, T]$,

$$\|\hat{i}_\eta T_\eta(t)u_\eta - T_0(t)\hat{i}_\eta u_\eta\|_{X_0^{1/2}} < \varepsilon \text{ and } \|\hat{j}_\eta T_0(t)u_0 - T_\eta(t)\hat{j}_\eta u_0\|_{X_\eta^{1/2}} < \varepsilon.$$

By Lemma 7.10, for any $T > 0$ and $\varepsilon > 0$, we can choose $\eta_1 > 0$ and $0 < \delta < \varepsilon/3$ such that if $\eta \in [0, \eta_1]$ and $\|u_\eta - v_\eta\|_{X_\eta^{1/2}} + |t - s| < \delta$ for $u_\eta, v_\eta \in \mathcal{A}_\eta$ and $t, s \in [-T, T]$, then

$$\|T_\eta(t)u_\eta - T_\eta(s)v_\eta\|_{X_\eta^{1/2}} < \frac{\varepsilon}{3}.$$

Since the set $\bigcup_{0 \leq \eta < \eta_1} i_\eta \mathcal{A}_\eta$ is precompact in $X_0^{1/2}$, we can assume that $\delta > 0$ is such that if $u, v \in \bigcup_{0 \leq \eta < \eta_1} i_\eta \mathcal{A}_\eta$ with $\|u - v\|_{X_0^{1/2}} < \delta$, then

$$\|T_0(t)u - T_0(t)v\|_{X_0^{1/2}} < \frac{\varepsilon}{9} \tag{7.8}$$

for all $t \in [-T, T]$. By Lemmas 7.4 and 7.10, we take $\eta_2 > 0$ such that

$$\|i_\eta T_\eta(t)j_\eta u_0 - T_0(t)u_0\|_{X_0^{1/2}} < \varepsilon/9 \tag{7.9}$$

for $u_0 \in \mathcal{A}_0$, $t \in [0, T]$ and $\eta \in [0, \eta_2]$. By Step 3 in Theorem 7.3, there is $0 < \eta_3 < \eta_2$ such that if $\eta \in [0, \eta_3]$, then there are $\delta/3$-isometries

$$\hat{i}_\eta : \mathcal{A}_\eta \to \mathcal{A}_0 \quad \text{and} \quad \hat{j}_\eta : \mathcal{A}_0 \to \mathcal{A}_\eta$$

satisfying

$$\|\hat{i}_\eta u_\eta - i_\eta u_\eta\|_{X_0^{1/2}} < \frac{\delta}{3} \quad \text{and} \quad \|\hat{j}_\eta u_0 - j_\eta u_0\|_{X_\eta^{1/2}} < \frac{\delta}{3} \tag{7.10}$$

for $u_\eta \in \mathcal{A}_\eta$ and $u_0 \in \mathcal{A}_0$. For any $\eta \in [0, \eta_3]$, $u_0 \in \mathcal{A}_0$, and $t \in [0, T]$, we have

$$\|\hat{i}_\eta T_\eta(t)\hat{j}_\eta u_0 - T_0(t)u_0\|_{X_0^{1/2}} \leq \|\hat{i}_\eta T_\eta(t)\hat{j}_\eta u_0 - i_\eta T_\eta(t)\hat{j}_\eta u_0\|_{X_0^{1/2}}$$
$$+ \|i_\eta T_\eta(t)\hat{j}_\eta u_0 - T_0(t)i_\eta \hat{j}_\eta u_0\|_{X_0^{1/2}}$$
$$+ \|T_0(t)i_\eta \hat{j}_\eta u_0 - T_0(t)u_0\|_{X_0^{1/2}} := I_1 + I_2 + I_3.$$

By (7.10), we get $I_1 < \varepsilon/9$. By (7.9), we have

$$I_2 = \|i_\eta T_\eta(t)j_\eta i_\eta \hat{j}_\eta u_0 - T_0(t)i_\eta \hat{j}_\eta u_0\|_{X_0^{1/2}} < \frac{\varepsilon}{9}.$$

Since

$$\|i_\eta \hat{j}_\eta u_0 - u_0\|_{X_0^{1/2}} \leq 2\|\hat{j}_\eta u_0 - j_\eta u_0\|_{X_0^{1/2}} \leq \frac{2\delta}{3},$$

by (7.8), we obtain $I_3 < \varepsilon/9$. Consequently, we deduce

$$\|\hat{i}_\eta T_\eta(t)\hat{j}_\eta u_0 - T_0(t)u_0\|_{X_0^{1/2}} < \frac{\varepsilon}{3}. \tag{7.11}$$

For $\eta \in [0, \eta_3]$ and $u_0 \in \mathcal{A}_0$, we have

$$\|\hat{i}_\eta \circ \hat{j}_\eta u_0 - u_0\|_{X_0^{1/2}} \leq \|\hat{i}_\eta \circ \hat{j}_\eta u_0 - i_\eta \circ \hat{j}_\eta u_0\|_{X_0^{1/2}} + \|i_\eta \circ \hat{j}_\eta u_0 - i_\eta \circ j_\eta u_0\|_{X_0^{1/2}}$$

$$\leq \frac{\delta}{3} + 2\|\hat{j}_\eta u - j_\eta u\|_{X_0^{1/2}} < \delta. \tag{7.12}$$

For $u_\eta \in \mathcal{A}_\eta$, we see that

$$\|\hat{j}_\eta \circ \hat{i}_\eta u_\eta - u_\eta\|_{X_0^{1/2}} \leq \|\hat{j}_\eta \circ \hat{i}_\eta u_\eta - j_\eta \circ \hat{i}_\eta u_\eta\|_{X_0^{1/2}}$$

$$+ \|j_\eta \circ \hat{i}_\eta u_\eta - j_\eta \circ i_\eta u_0\|_{X_0^{1/2}}$$

$$\leq \frac{\delta}{3} + 2\|\hat{i}_\eta u - i_\eta u\|_{X_0^{1/2}} < \delta. \tag{7.13}$$

For any $u_\eta \in \mathcal{A}_\eta$ and $t \in [0, T]$, we have

$$\|\hat{i}_\eta T_\eta(t) u_\eta - T_0(t) \hat{i}_\eta u_\eta\|_{X_0^{1/2}}$$

$$\leq \|\hat{i}_\eta T_\eta(t) u_\eta - \hat{i}_\eta T_\eta(t) \hat{j}_\eta \hat{i}_\eta u_\eta\|_{X_0^{1/2}} + \|\hat{i}_\eta T_\eta(t) \hat{j}_\eta \hat{i}_\eta u_\eta - T_0(t) \hat{i}_\eta u_\eta\|_{X_0^{1/2}}$$

$$\leq \frac{\varepsilon}{3} + \|T_\eta(t) u_\eta - T_\eta(t) \hat{j}_\eta \hat{i}_\eta u_\eta\|_{X_0^{1/2}} + \|\hat{i}_\eta T_\eta(t) \hat{j}_\eta \hat{i}_\eta u_\eta - T_0(t) \hat{i}_\eta u_\eta\|_{X_0^{1/2}}$$

$$= \frac{\varepsilon}{3} + II_1 + II_2.$$

By (7.13) and Lemma 7.10, $II_1 < \varepsilon/3$. By (7.11), $II_2 < \varepsilon/3$. Consequently,

$$\|\hat{i}_\eta T_\eta(t) u_\eta - T_0(t) \hat{i}_\eta u_\eta\|_{X_0^{1/2}} < \varepsilon.$$

For any $u_0 \in \mathcal{A}_0$ and $t \in [0, T]$, we have

$$\|\hat{j}_\eta T_0(t) u_0 - T_\eta(t) \hat{j}_\eta u_0\|_{X_0^{1/2}} \leq \|\hat{i}_\eta \circ \hat{j}_\eta T_0(t) u_0 - \hat{i}_\eta T_\eta(t) \hat{j}_\eta u_0\|_{X_0^{1/2}} + \varepsilon/3$$

$$\leq \|\hat{i}_\eta \circ \hat{j}_\eta T_0(t) u_0 - T_0(t) u_0\|_{X_0^{1/2}}$$

$$+ \|T_0(t) u_0 - \hat{i}_\eta T_\eta(t) \hat{j}_\eta u_0\|_{X_0^{1/2}} + \varepsilon/3$$

$$= III_1 + III_2 + \varepsilon/3.$$

By (7.12), $III_1 < \varepsilon/3$. By (7.11), $III_2 < \varepsilon/3$. It follows that

$$\|\hat{j}_\eta T_0(t) u_0 - T_\eta(t) \hat{j}_\eta u_0\|_{X_0^{1/2}} < \varepsilon.$$

Step 2. We show that the dynamical system T_0 on $\mathcal{A}_0$ is Gromov-Hausdorff stable.

For $T > 0$ and $\varepsilon > 0$, by Lemma 7.10, there are $\eta_4 > 0$ and $0 < \gamma < \varepsilon/2$ such that if $\eta \in [0, \eta_4]$ and $\|u_\eta - v_\eta\|_{X_\eta^{1/2}} + |t - s| < \gamma$ for $u_\eta, v_\eta \in \mathcal{A}_\eta$ and $t, s \in [-2T, 2T]$, then

$$\|T_\eta(t)u_\eta - T_\eta(s)v_\eta\|_{X_\eta^{1/2}} < \frac{\varepsilon}{2}. \tag{7.14}$$

By Step 1, we can take $0 < \eta_0 < \eta_4$ such that for any $\eta \in [0, \eta_0]$, there exist γ-isometries

$$\hat{i}_\eta : \mathcal{A}_\eta \to \mathcal{A}_0 \quad \text{and} \quad \hat{j}_\eta : \mathcal{A}_0 \to \mathcal{A}_\eta$$

such that for any $u_0 \in \mathcal{A}_0$, $u_\eta \in \mathcal{A}_\eta$, and $t \in [0, 2T]$,

$$\|\hat{i}_\eta T_\eta(t)u_\eta - T_0(t)\hat{i}_\eta u_\eta\|_{X_0^{1/2}} < \gamma \quad \text{and} \quad \|\hat{j}_\eta T_0(t)u_0 - T_\eta(t)\hat{j}_\eta u_0\|_{X_\eta^{1/2}} < \gamma. \tag{7.15}$$

For any $\eta \in [0, \eta_0]$, $u_0 \in \mathcal{A}_0$, and $t \in [-T, 0]$, we take $w_0 \in \mathcal{A}_0$ such that $u_0 = T_0(-2t)w_0$. Then we have

$$\|\hat{j}_\eta T_0(t)u_0 - T_\eta(t)\hat{j}_\eta u_0\|_{X_\eta^{1/2}} = \|\hat{j}_\eta T_0(-t)w_0 - T_\eta(t)\hat{j}_\eta T_0(-2t)w_0\|_{X_\eta^{1/2}}$$

$$\leq \|\hat{j}_\eta T_0(-t)w_0 - T_\eta(-t)\hat{j}_\eta w_0\|_{X_\eta^{1/2}}$$

$$+ \|T_\eta(t)T_\eta(-2t)\hat{j}_\eta w_0 - T_\eta(t)\hat{j}_\eta T_0(-2t)w_0\|_{X_\eta^{1/2}}$$

$$= I + II.$$

By (7.15), we have $I < \varepsilon/2$ and

$$\|T_\eta(-2t)\hat{j}_\eta w_0 - \hat{j}_\eta T_0(-2t)w_0\|_{X_\eta^{1/2}} \leq \gamma.$$

Then, by (7.14), we get $II < \varepsilon/2$. So, $\|\hat{j}_\eta T_0(t)u_0 - T_\eta(t)\hat{j}_\eta u_0\|_{X_\eta^{1/2}} < \varepsilon$.

For any $\eta \in [0, \eta_0]$, $u_\eta \in \mathcal{A}_\eta$, and $t \in [-T, 0]$, we take $w_\eta \in \mathcal{A}_\eta$ such that $u_\eta = T_\eta(-2t)w_\eta$. By (7.14) and (7.15), we have

$$\|\hat{i}_\eta T_\eta(t)u_\eta - T_0(t)\hat{i}_\eta u_\eta\|_{X_0^{1/2}} = \|\hat{i}_\eta T_\eta(-t)w_\eta - T_0(t)\hat{i}_\eta T_\eta(-2t)w_\eta\|_{X_0^{1/2}}$$

$$\leq \|\hat{i}_\eta T_\eta(-t)w_\eta - T_0(-t)\hat{i}_\eta w_\eta\|_{X_0^{1/2}}$$

$$+ \|T_0(t)T_0(-2t)\hat{i}_\eta w_\eta - T_0(t)\hat{i}_\eta T_\eta(-2t)w_\eta\|_{X_0^{1/2}}$$

$$\leq \varepsilon.$$

This completes the proof of Theorem 7.5. $\qquad\qquad\qquad\qquad\qquad\qquad\qquad\square$

Now we are ready to prove the main theorem of in this section, which asserts that the system (7.1) is geometrically stable under Lipschitz perturbations of the domain and equation. To prove this, we use the shadowing property of (7.1) on the global attractor $\mathcal{A}_0$, which was proved by S. Pilyugin.

Let T be a homeomorphism on a compact metric space (X, d). A sequence $\xi = \{x_n \in X : a < n < b\}$ ($-\infty \leq a < b \leq \infty$) is called a δ-pseudo-orbit of T if $d(T(x_n), x_{n+1}) < \delta$ for all $a < n < b - 1$. We say that a δ-pseudo-orbit $\xi = \{x_n \in X : a < n < b\}$ of T is ε-shadowed by a point $x \in X$ if $d(T^n(x), x_n) < \varepsilon$ for all $a < n < b$; in this case, the point x is called a shadowing point of the pseudo-orbit ξ. By Theorem 3.4.1 in [62], we can derive that the time t-map $T_0(t)$ ($t > 0$) induced by (7.1) has the forward shadowing property, that is, for any $\varepsilon > 0$, there is $\delta > 0$ such that any δ-pseudo-orbit $\xi = \{x_n\}_{n \in \mathbb{N}}$ of $T_0(t)$ in its global attractor $\mathcal{A}_0$ can be ε-shadowed by a point $x \in \mathcal{M}_0$. Note that the shadowing point x does not necessarily belong to $\mathcal{A}_0$.

In the following lemma, we prove that the time t-map $T_0(t)$ has the intrinsic shadowing property on $\mathcal{A}_0$, that is, for any $\varepsilon > 0$, there is $\delta > 0$ such that any δ-pseudo-orbit $\xi = \{x_n\}_{n \in \mathbb{Z}}$ in $\mathcal{A}_0$ can be ε-shadowed by a point $x \in \mathcal{A}_0$.

Lemma 7.11 (Intrinsic Shadowing Property) *For any $t > 0$, the time t-map $T_0(t)$ of T_0 has the intrinsic shadowing property on the global attractor $\mathcal{A}_0$.*

Proof For $t > 0$, we see that the time t-map $T_0(t)$ satisfies the assumption of Theorem 3.4.1 in [62]. Then there are $d_0, L_0 > 0$ and a neighborhood W of $\mathcal{A}_0$ in $X_0^{1/2}$ such that any d-pseudo-orbit $\xi = \{u_k\}_{k \geq 0} \subset W$ ($d < d_0$) for which $\mathrm{dist}(u_0, \mathcal{M}_0) < d$ can be $L_0 d$-shadowed by a point $u \in \mathcal{M}_0$, that is, for any $k \geq 0$,

$$\|T_0(kt)u - u_k\|_{X_0^{1/2}} \leq L_0 d.$$

For $\varepsilon > 0$, we take $\delta = \min\{\frac{d_0}{2}, \frac{\varepsilon}{L_0}\}$. Let $\xi = \{u_k\}_{k \in \mathbb{Z}}$ be a δ-pseudo-orbit of $T_0(t)$ in $\mathcal{A}_0$. For any $n \in \mathbb{N}$, the δ-pseudo orbit $\xi_n = \{u_k\}_{k \geq -n}$ can be ε-shadowed by $v_n \in \mathcal{M}_0$, that is, for any $k \geq 0$,

$$\|T_0(kt)v_n - u_{k-n}\|_{X_0^{1/2}} \leq \varepsilon.$$

Since $\{T_0(nt)v_n\}_{n \in \mathbb{N}}$ is bounded in the inertial manifold $\mathcal{M}_0$, there is a subsequence $\{T_0(n_i t)v_{n_i}\}$ such that $T_0(n_i t)v_{n_i} \to v$ for some $v \in \mathcal{M}_0$. For any $k \in \mathbb{Z}$, we have

$$\|T_0(kt)v - u_k\|_{X_0^{1/2}} = \lim_{i \to \infty} \|T_0(kt)T_0(n_i t)v_{n_i} - u_k\|_{X_0^{1/2}} < \varepsilon.$$

Then v is a shadowing point of ξ. Since ξ is bounded in $X_0^{1/2}$, $\{T_0(kt)v\}$ is bounded in $X_0^{1/2}$, and so $v \in \mathcal{A}_0$. $\square$

In view of Lemma 7.8, if every equilibrium of $T_0(1)$ is hyperbolic, then there is $\eta_0 > 0$ such that if $\eta \in [0, \eta_0]$, then $\tilde{T}_\eta(1)$ has the same number of equilibria that are $\mathcal{L}$-hyperbolic. Then by Bortolan et al. [13, Proposition 3.9, Theorem 3.10], we can take a uniform $r > 0$ such that any equilibrium $u^*_{i,\eta}$ of $T_\eta(1)$ is isolated in its r-neighborhood, that is, if $\{T_\eta(n)u_\eta\}_{n \in \mathbb{Z}}$ is a bounded global solution of T_η, then

$$\|T_\eta(n)u_\eta - u^*_{i,\eta}\|_{X^{1/2}_\eta} < r \text{ for } n \in \mathbb{Z} \quad \text{implies} \quad u_\eta = u^*_{i,\eta}. \tag{7.16}$$

Proof (Proof of Theorem 7.4) Let $r > 0$ be a constant in (7.16). For any $0 < \varepsilon < r/2$, take $\eta_0 > 0$ and $0 < \delta < 1$ corresponding to $\varepsilon/4$ and $T = 1$ as in Lemma 7.10. Also, choose $0 < \gamma < \varepsilon/4$ such that any γ-pseudo-orbit of $T_0(\delta)$ in $\mathcal{A}_0$ can be $\varepsilon/4$-shadowed by a point in $\mathcal{A}_0$. By Theorem 7.5, we can choose a smaller $\eta_0 > 0$ if necessary, such that for any $\eta \in [0, \eta_0]$, there is a γ-isometry $\hat{i}_\eta : \mathcal{A}_\eta \to \mathcal{A}_0$ such that for $u_\eta \in \mathcal{A}_\eta$ and $t \in [-\delta, \delta]$,

$$\|\hat{i}_\eta T_\eta(t)u_\eta - T_0(t)\hat{i}_\eta u_\eta\|_{X^{1/2}_0} < \gamma.$$

For $\eta \in [0, \eta_0]$, we take a subset $\mathcal{D}_\eta$ of $\mathcal{A}_\eta$ such that

(i) for $v_\eta \in \mathcal{A}_\eta$, there is $u_\eta \in \mathcal{D}_\eta$ such that $v_\eta = T_\eta(t)u_\eta$ for some $t \in \mathbb{R}$;
(ii) if $u_\eta, v_\eta \in \mathcal{D}_\eta$ and $v_\eta = T_\eta(t)u_\eta$ for some $t \in \mathbb{R}$, then $u_\eta = v_\eta$.

For $u_\eta \in \mathcal{D}_\eta$, consider a sequence $\{\hat{i}_\eta T_\eta(n\delta)u_\eta\}_{n \in \mathbb{Z}}$ in $\mathcal{A}_0$. For any $n \in \mathbb{Z}$, we have

$$\|T_0(\delta)\hat{i}_\eta T_\eta(n\delta)u_\eta - \hat{i}_\eta T_\eta(\delta)T_\eta(n\delta)u_\eta\|_{X^{1/2}_0} < \gamma.$$

Then $\{\hat{i}_\eta T_\eta(n\delta)u_\eta\}_{n \in \mathbb{Z}}$ is a γ-pseudo-orbit of $T_0(\delta)$. By Lemma 7.11, there is an element in $\mathcal{A}_0$, denoted by $\tilde{i}_\eta(u_\eta)$, such that

$$\|T_0(n\delta)\tilde{i}_\eta u_\eta - \hat{i}_\eta T_\eta(n\delta)u_\eta\|_{X^{1/2}_0} < \frac{\varepsilon}{4}.$$

For any $t \in \mathbb{R}$, we can write $t = n\delta + s$ for some $n \in \mathbb{Z}$ and $0 \leq s < \delta$. We have

$$\|T_0(t)\tilde{i}_\eta u_\eta - \hat{i}_\eta T_\eta(t)u_\eta\|_{X^{1/2}_0}$$

$$\leq \|T_0(s)T_0(n\delta)\tilde{i}_\eta u_\eta - T_0(n\delta)\tilde{i}_\eta u_\eta\|_{X^{1/2}_0} +$$

$$\|T_0(n\delta)\tilde{i}_\eta u_\eta - \hat{i}_\eta T_\eta(n\delta)u_\eta\|_{X^{1/2}_0} + \|\hat{i}_\eta T_\eta(n\delta)u_\eta - \hat{i}_\eta T_\eta(s)T_\eta(n\delta)u_\eta\|_{X^{1/2}_0}$$

$$< \varepsilon.$$

We define a map $\tilde{i}_\eta : \mathcal{D}_\eta \to \mathcal{A}_0$ as follows: for any $u_\eta \in \mathcal{D}_\eta$, $\tilde{i}_\eta u_\eta \in \mathcal{A}_0$ is a shadowing point of the γ-pseudo orbit $\{\hat{i}_\eta T_\eta(n\delta)u_\eta\}_{n\in\mathbb{Z}}$. For any $u_\eta \in \mathcal{A}_\eta$, choose $v_\eta \in \mathcal{D}_\eta$ such that $u_\eta = T_\eta(t)v_\eta$ for some $t \in \mathbb{R}$. We define $\tilde{i}_\eta(u_\eta) = T_0(t)\tilde{i}_\eta v_\eta$.

We observe that $\tilde{i}_\eta$ is an ε-isometry. In fact, for any $u_\eta, v_\eta \in \mathcal{A}_\eta$, there are $\tilde{u}_\eta, \tilde{v}_\eta \in \mathcal{D}_\eta$ such that $u_\eta = T_\eta(t)\tilde{u}_\eta$ and $v_\eta = T_\eta(s)\tilde{v}_\eta$ for some $t, s \in \mathbb{R}$. Then we have

$$\|\tilde{i}_\eta u_\eta - \tilde{i}_\eta v_\eta\|_{X_0^{1/2}} - \|u_\eta - v_\eta\|_{X_\eta^{1/2}}$$

$$\leq \|\tilde{i}_\eta T_\eta(t)\tilde{u}_\eta - \hat{i}_\eta T_\eta(t)\tilde{u}_\eta\|_{X_0^{1/2}} + \|\hat{i}_\eta T_\eta(t)\tilde{u}_\eta - \hat{i}_\eta T_\eta(s)\tilde{v}_\eta\|_{X_0^{1/2}}$$

$$+ \|\hat{i}_\eta T_\eta(s)\tilde{v}_\eta - \tilde{i}_\eta T_\eta(s)\tilde{v}_\eta\|_{X_0^{1/2}} - \|u_\eta - v_\eta\|_{X_\eta^{1/2}}$$

$$:= I + II + III - \|u_\eta - v_\eta\|_{X_\eta^{1/2}}.$$

By the shadowing property of T_0 on $\mathcal{A}_0$,

$$I = \|T_0(t)\tilde{i}_\eta\tilde{u}_\eta - \hat{i}_\eta T_\eta(t)\tilde{u}_\eta\|_{X_0^{1/2}} < \frac{\varepsilon}{4}$$

and

$$III = \|T_0(s)\tilde{i}_\eta\tilde{v}_\eta - \hat{i}_\eta T_\eta(s)\tilde{v}_\eta\|_{X_0^{1/2}} < \frac{\varepsilon}{4}.$$

Moreover, we get

$$II = \|\hat{i}_\eta T_\eta(t)\tilde{u}_\eta - \hat{i}_\eta T_\eta(s)\tilde{v}_\eta\|_{X_0^{1/2}} \leq \frac{\varepsilon}{4} + \|u_\eta - v_\eta\|_{X_\eta^{1/2}}.$$

Hence,

$$\|\tilde{i}_\eta u_\eta - \tilde{i}_\eta v_\eta\|_{X_0^{1/2}} - \|u_\eta - v_\eta\|_{X_\eta^{1/2}} < \varepsilon.$$

Similarly, we have

$$\|u_\eta - v_\eta\|_{X_\eta^{1/2}} - \|\tilde{i}_\eta u_\eta - \tilde{i}_\eta v_\eta\|_{X_0^{1/2}} < \varepsilon,$$

and so

$$\left| \|\tilde{i}_\eta u_\eta - \tilde{i}_\eta v_\eta\|_{X_0^{1/2}} - \|u_\eta - v_\eta\|_{X_\eta^{1/2}} \right| < \varepsilon.$$

Moreover, for any $u_0 \in \mathcal{A}_0$, there is $u_\eta \in \mathcal{A}_\eta$ such that $\|u_0 - \hat{i}_\eta u_\eta\|_{X_0^{1/2}} < \varepsilon/4$. Then

$$\|\tilde{i}_\eta u_\eta - u_0\|_{X_0^{1/2}} \le \|\tilde{i}_\eta u_\eta - \hat{i}_\eta u_\eta\|_{X_0^{1/2}} + \|\hat{i}_\eta u_\eta - u_0\|_{X_0^{1/2}} < \varepsilon.$$

Consequently, $\mathcal{A}_0 \subset B(\tilde{i}_\eta \mathcal{A}_\eta, \varepsilon)$, and so $\tilde{i}_\eta$ is an ε-isometry.

For any $u_\eta \in \mathcal{A}_\eta$ and $t \in \mathbb{R}$, there is $v_\eta \in \mathcal{D}_\eta$ such that $T_\eta(s) v_\eta = u_\eta$ for some $s \in \mathbb{R}$. Then we have

$$\tilde{i}_\eta T_\eta(t) u_\eta = \tilde{i}_\eta T_\eta(t) T_\eta(s) v_\eta = T_0(t) T_0(s) \tilde{i}_\eta v_\eta = T_0(t) \tilde{i}_\eta T_\eta(s) v_\eta = T_0(t) \tilde{i}_\eta u_\eta.$$

Finally, we show that $\tilde{i}_\eta|_{\mathcal{E}_\eta} : \mathcal{E}_\eta \to \mathcal{E}_0$ is a bijection and

$$u_\eta \in W^u(u_{i,\eta}^*, T_\eta(1)) \cap W^s_{\text{loc}}(u_{j,\eta}^*, T_\eta(1))$$

if and only if

$$\tilde{i}_\eta u_\eta \in W^u(\tilde{i}_\eta u_{i,\eta}^*, T_0(1)) \cap W^s_{\text{loc}}(\tilde{i}_\eta u_{j,\eta}^*, T_0(1)),$$

where $1 \le i, j \le p$. Note that $\mathcal{E}_\eta \subset \mathcal{D}_\eta$. Then we have

$$\|T_0(n) \tilde{i}_\eta u_{i,\eta}^* - \hat{i}_\eta u_{i,\eta}^*\|_{X_0^{1/2}} < \frac{\varepsilon}{4}$$

for $n \in \mathbb{Z}$. Hence, if $\eta > 0$ is sufficiently small, we get

$$\|T_0(n) \tilde{i}_\eta u_{i,\eta}^* - u_{i,0}^*\|_{X_0^{1/2}} \le \|T_0(n) \tilde{i}_\eta u_{i,\eta}^* - \hat{i}_\eta u_{i,\eta}^*\|_{X_0^{1/2}}$$
$$+ \|\hat{i}_\eta u_{i,\eta}^* - i_\eta u_{i,\eta}^*\|_{X_0^{1/2}} + \|i_\eta u_{i,\eta}^* - u_{i,0}^*\|_{X_0^{1/2}} < r.$$

Since any equilibrium of $T_0(1)$ is isolated, we conclude that $\tilde{i}_\eta u_{i,\eta}^* = u_{i,0}^*$.

Let $u_\eta \in W^u(u_{i,\eta}^*, T_\eta(1)) \cap W^s_{\text{loc}}(u_{j,\eta}^*, T_\eta(1))$. Consider a bounded global solution $\xi : \mathbb{Z} \to \mathcal{A}_\eta$ given by $\xi(n) = T_\eta(n) u_\eta$ for any $n \in \mathbb{Z}$. Take $N > 0$ such that

$$\|\xi(n) - u_{i,\eta}^*\|_{X_\eta^{1/2}} < r/2 \text{ for } n \le -N \quad \text{and} \quad \|\xi(n) - u_{j,\eta}^*\|_{X_\eta^{1/2}} < r/2 \text{ for } n \ge N.$$

Then we have

$$\|T_0(n) \tilde{i}_\eta u_\eta - u_{i,0}^*\|_{X_0^{1/2}} = \|\tilde{i}_\eta T_\eta(n) u_\eta - \tilde{i}_\eta u_{i,\eta}^*\|_{X_0^{1/2}} \le r.$$

In much the same way,

$$\|T_0(n) \tilde{i}_\eta u_\eta - u_{j,0}^*\|_{X_0^{1/2}} < r.$$

Since $T_0(1)$ is dynamically gradient with respect to $\mathcal{E}_0$, we have

$$T_0(n)\tilde{i}_\eta u_\eta \to u^*_{i,0} \text{ as } n \to -\infty \quad \text{and} \quad T_0(n)\tilde{i}_\eta u_\eta \to u^*_{j,0} \text{ as } n \to \infty.$$

This implies that $\tilde{i}_\eta u_\eta \in W^u(\tilde{i}_\eta u^*_{i,\eta}, T_0(1)) \cap W^s_{loc}(\tilde{i}_\eta u^*_{j,\eta}, T_0(1))$.

Conversely, for $\tilde{i}_\eta u_\eta \in W^u(\tilde{i}_\eta u^*_{i,\eta}, T_0(1)) \cap W^s_{loc}(\tilde{i}_\eta u^*_{j,\eta}, T_0(1))$, we have

$$T_0(n)\tilde{i}_\eta u_\eta \to \tilde{i}_\eta u^*_{i,\eta} \text{ as } n \to -\infty \quad \text{and} \quad T_0(n)\tilde{i}_\eta u_\eta \to \tilde{i}_\eta u^*_{j,\eta} \text{ as } n \to \infty.$$

Hence, there is $M > 0$ such that

$$\|T_0(n)\tilde{i}_\eta u_\eta - \tilde{i}_\eta u^*_{i,\eta}\|_{X_0^{1/2}} < \frac{r}{2} \quad \text{and} \quad \|T_0(n)\tilde{i}_\eta u_\eta - \tilde{i}_\eta u^*_{j,\eta}\|_{X_0^{1/2}} < \frac{r}{2}$$

for $n \le -M$ and $n \ge M$, respectively. Then for any $n \le -M$,

$$\|T_\eta(n)u_\eta - u^*_{i,\eta}\|_{X_\eta^{1/2}} \le \varepsilon + \|\tilde{i}_\eta T_\eta(n)u_\eta - \tilde{i}_\eta u^*_{i,\eta}\|_{X_0^{1/2}}$$

$$= \varepsilon + \|T_0(n)\tilde{i}_\eta u_\eta - u^*_{i,0}\|_{X_0^{1/2}}$$

$$< r,$$

and for any $n \ge M$,

$$\|T_\eta(n)u_\eta - u^*_{j,\eta}\|_{X_\eta^{1/2}} < r.$$

Since $T_\eta(1)$ is dynamically gradient with respect to $\mathcal{E}_\eta$,

$$T_\eta(n)u_\eta \to u^*_{i,\eta} \text{ as } n \to -\infty \quad \text{and} \quad T_\eta(n)u_\eta \to u^*_{j,\eta} \text{ as } n \to \infty.$$

Consequently, $u_\eta \in W^u(u^*_{i,\eta}, T_\eta(1)) \cap W^s_{loc}(u^*_{j,\eta}, T_\eta(1))$. This completes the proof.   □

References

1. Andronov, A.A., Pontryagin, L.S.: Systémes grossiers. Dokl. Akad. Nauk. SSSR. **14**, 247–251 (1937)
2. Anosov, D.V.: Roughness of geodesic flows on compact Riemannian manifolds of negative curvature. Dokl. Akad. Nauk SSSR 145, 707–709 (1962)
3. Aoki, N., Hiraide, K.: Topological Theory of Dynamical Systems. Recent Advances, North-Holland Mathematical Library, vol. 52. North-Holland Publishing Co., Amsterdam (1994)
4. Aragão-Costa, E.R., Figueroa-López, R.N., Langa, J.A., Lozada-Cruz, G.: Topological structural stability of partial differential equations on projected spaces. J. Dynam. Differential Equations **30**, 687–718 (2018)
5. Arbieto, A., Morales, C.A.: Topological stability from Gromov-Hausdorff viewpoint. Discrete Contin. Dyn. Syst. **37**, 3531–3544 (2017)
6. Arrieta, J.M., Carvalho, A.N.: Spectral convergence and nonlinear dynamics of reaction-diffusion equations under perturbations of the domain, J. Differential Equations **199**, 143–178 (2004)
7. Arrieta, J.M., Carvalho, A.N., Rodriguez-Bernal, A.: Attractors for parabolic problems with nonlinear boundary condition. Uniform bounds. Commun. Partial Differential Equations **25**, 1–37 (2000)
8. Arrieta, J.M., Santamaria, E.: Estimates on the distance of inertial manifolds. Discrete Conti. Dyn. Syst. **34**, 3921–3944 (2014)
9. Artigue, A.: Lipschitz perturbations of expansive systems. Discrete Contin. Dyn. Syst. **35**, 1829–1841 (2015)
10. Artigue, A.: Generic dynamics on compact metric spaces. Topology Appl. **255**, 1–14 (2019)
11. Babin, A.V., Pilyugin, S.Y.: Continuous dependence of an attractor on the shape of domain. Zap. Nauchn. Sem. POMI **221**, 58–66 (1995). J. Math. Sci. **87**, 3304–3310 (1997)
12. Barzanouni, A.: Inverse limit spaces with various shadowing property. J. Math. **2014**, Art. ID 169183, 4 pp. (2014)
13. Bortolan, M.C., Cardoso, C.A.E.N., Carvalho, A.N., Pires, L.: Lipschitz perturbations of Morse-Smale semigroups. J. Differential Equations **269**, 1904–1943 (2020)
14. Burago, D., Burago, Y., Ivanov, S.: A Course in Metric Geometry, Graduate Studies in Mathematics, vol. 33. American Mathematical Society, Providence (2001)
15. Chafee, N., Infante, E.F.: A bifurcation problem for a nonlinear partial differential equation of parabolic type. Appl. Anal. **4**, 17–37 (1974)
16. Cheeger, J., Fukaya K., Gromov M.: Nilpotent structures and invariant metrics on collapsed manifolds. J. Amer. Math. Soc. **6**, 327–372 (1992)

© The Author(s), under exclusive license to Springer Nature Switzerland AG 2022
J. Lee, C. Morales, *Gromov-Hausdorff Stability of Dynamical Systems and Applications to PDEs*, Frontiers in Mathematics, https://doi.org/10.1007/978-3-031-12031-2

17. Chen, L., Li, S.H.: Shadowing property for inverse limit spaces. Proc. Amer. Math. Soc. **115**, 573–580 (1992)
18. Chen, Z.P., Liu, P.D.: Orbit-shift stability of class of self-covering maps, Sci. China Ser. A **34**(1), 1–13 (1991)
19. Chen, Z.P., He, L.F., Liu, P.D.: Orbit-shift topological Ω-stability. Acta Math. Sin. **32**, 71–75 (1989)
20. Chepyzhov, V.V., Vishik, M.I.: Attractors for Equations of Mathematical Physics, American Mathematical Society Colloquium Publications, vol. 49. American Mathematical Society, Providence (2002)
21. Chulluncuy, A.: Topological stability for flows from a Gromov-Hausdorff viewpoint. Bull. Braz. Math. Soc. (N.S.) **53**, 307–341 (2022)
22. Chung, N.P.: Gromov-Hausdorff distances for dynamical systems. Discrete Contin. Dyn. Syst. **40**, 6179–6200 (2020)
23. Chung, N.P., Lee, K.: Topological stability and pseudo-orbit tracing property of group actions. Proc. Amer. Math. Soc. **146**, 1047–1057 (2018)
24. Cubas, R.: Properties of a GH-stable homeomorphism, M.Sc. Dissertation, Federal University of Uberlandia, Uberlandia-MG, 2018, p. 51
25. Dong, M.: Group actions from measure theoretical viewpoint, PhD. Thesis, Chungnam National University, Daejeon, 2018
26. Edrei, A.: On mappings which do not increase small distances. Proc. London Math. Soc. (3) **2**, 272–278 (1952)
27. Edwards, D.A.: The structure of superspace. In: Studies in Topology (Proc. Conf., Univ. North Carolina, Charlotte, N. C., 1974; dedicated to Math. Sect. Polish Acad. Sci.), pp. 121–133. Academic Press, New York (1975)
28. Foias, C., Sell, G., Temam, R.: Inertial manifolds for nonlinear evolutionary equations. J. Differential Equations **73**, 309–353 (1988)
29. Fukaya, K.: Collapsing of Riemannian manifolds and eigenvalues of Laplace operator. Invent. Math. **87**, 517–547 (1987)
30. Fukaya K.: Hausdorff convergence of Riemannian manifolds and its applications. In: Recent Topics in Differential and Analytic Geometry, vol. 18, pp. 143–238. Academic Press, Boston (1990)
31. Gromov, M.: Groups of polynomial growth and expanding maps. Inst. Hautes Études Sci. Publ. Math. **53**, 53–73 (1981)
32. Hale, J.K.: Topics in Dynamic Bifurcation Theory, CBMS Lecture Notes, vol. 47. Amer. Math. Soc. Providence (1981)
33. Hale, J.K., Magalhães, L.T., Olivia, W.M.: Dynamics in Infinite Dimensions, 2nd edn. Appl. Math. Sci., vol. 47 (Springer, 2002)
34. Henry, D.B.: Geometric Theory of Semilinear Parabolic Equations, Lecture Notes in Mathematics, vol. 840 (Springer, Berlin, 1981)
35. Henry, D.B.: Some infinite-dimensional Morse-Smale systems defined by parabolic partial differential equations. J. Differential Equations **59**, 165–205 (1985)
36. Henry, D.B.: Perturbation of the Boundary for Boundary Value Problems (Cambridge University Press, Cambridge, 2005)
37. Hoang, L.T., Olson, E.J., Robinson, J.C.: On the continuity of global attractors. Proc. Amer. Math. Soc. **143**, 4389–4395 (2015)
38. Ivanov, A.O., Tuzhilin, A.A.: Local structure of Gromov-Hausdorff space around generic finite metric spaces. Lobachevskii J. Math. **38**, 998–1006 (2017)
39. Ivanov, A.O., Nikolaeva, N.K., Tuzhilin, A.A.: The Gromov-Hausdorff metric on the space of compact metric spaces is strictly intrinsic. Math. Notes **100**, 883–885 (2016)

40. Jung, W.: The closure of periodic orbits in the Gromov-Hausdorff space. Topol. Appl. **264**, 493–497 (2019)
41. Kato, T.: Perturbation Theory for Linear Operators. Reprint of the 1980 edition. Classics in Mathematics. Springer, Berlin (1995)
42. Kostianko, A., Zelik, S.: Kwak transform and inertial manifolds. J. Dynam. Differential Equations. https://doi.org/10.1007/s10884-020-09913-9
43. Latremoliere, F.: Convergence of Cauchy sequences for the covariant Gromov-Hausdorff propinquity. J. Math. Anal. Appl. **469**, 378–404 (2019)
44. Lee, J.: Gromov-Hausdorff stability of reaction diffusion equations with Neumann boundary conditions under perturbations of the domain. J. Math. Anal. Appl. **496**, 124788 (2021)
45. Lee, J.: Stability of inertial manifolds for semilinear parabolic equations under Lipschitz perturbations, preprint
46. Lee, J., Nguyen, N.: Gromov-Hausdorff stability of inertial manifolds under perturbations of the domain and equation. J. Math. Anal. Appl. **494**, 124623 (2021)
47. Lee, J., Nguyen, N.: Topological stability of Chafee-Infante equations under Lipschitz perturbations of the domain and equation. J. Math. Anal. Appl. **517**, 126628 (2023)
48. Lee, J., Nguyen, T.: Gromov-Hausdorff stability of reaction diffusion equations with Robin boundary conditions under perturbations of the domain and equation. Comm. Pure and Appl. Anal. **20**, 1263–1296 (2021)
49. Lee, J., Nguyen, N., Toi, V.M.: Gromov-Hausdorff stability of global attractors of reaction diffusion equations under perturbations of the domain. J. Differential Equations **269**, 125–147 (2020)
50. Lewowicz, J.: Persistence in expansive systems. Ergodic Theory Dynam. Syst. **3**, 567–578 (1983)
51. Lu, K.: Structural stability for scalar parabolic equations. J. Differential Equations **114**, 253–271 (1994)
52. Lyapunov, A.M.: The General Problem of the Stability of Motion. Translated from Edouard Davaux's French translation (1907) of the 1892 Russian original and edited by A. T. Fuller. Reprint of Internat. J. Control 55, no. 3 (1992). With a foreword by Ian Stewart. Taylor & Francis, Ltd., London, 1992
53. Mallet-Paret, J., Sell, G.: Inertial manifolds for reaction diffusion equations in higher space dimensions. J. Amer. Math. Soc. **1**, 805–866 (1988)
54. Marion, M.: Attractors for reaction-diffusion equations: existence and estimate of their dimension. Appl. Anal. **25**, 101–147 (1987)
55. Nitecki, Z.: Differentiable Dynamics. An introduction to the Orbit Structure of Diffeomorphisms. The MIT Press, Cambridge (1971)
56. Nitecki, Z.: On semi-stability for diffeomorphisms. Invent. Math. **14**, 83–122 (1971)
57. Osipov, A.V., Tikhomirov, S.B.: Shadowing for actions of some finitely generated groups. Dyn. Syst. **29**, 337–351 (2014)
58. Palis, J., de Melo, W.: Geometric Theory of Dynamical Systems. Springer (1982)
59. Paluszynski, M., Stempak, K.: On quasi-metric and metric spaces. Proc. Amer. Math. Soc. **137**, 4307–4312 (2009)
60. Pereira, A.L., Pereira, M.C.: Continuity of attractors for a reaction-diffusion problem with nonlinear boundary conditions with respect to variations of the domain. J. Differential Equations **239**, 343–370 (2007)
61. Petersen, P.: Riemannian Geometry, 3rd edn. Graduate Texts in Mathematics, vol. 171. Springer, Cham (2016)
62. Pilyugin, S.Y.: Shadowing in Dynamical Systems, Lecture Notes in Math., vol. 1706. Springer (1999)

63. Pilyugin, S.Y., Tikhomirov, S.B.: Shadowing in actions of some Abelian groups. Fund. Math. **179**, 83–96 (2003)
64. Robinson, J.C.: Infinite-Dimensional Dynamical Systems. An Introduction to Dissipative Parabolic PDEs and the Theory of Global Attractors. Cambridge Texts in Applied Mathematics. Cambridge University Press, Cambridge (2001)
65. Robinson, J.C., Glendinning, P. (eds.): From Finite to Infinite Dimensional Dynamical Systems. Proceedings of the NATO Advanced Study Institute held in Cambridge, August 21–September 1, 1995. NATO Science Series II: Mathematics, Physics and Chemistry, vol. 19. Kluwer Academic Publishers, Dordrecht (2001)
66. Romanov, A.V.: Dimension of the central manifold for semilinear parabolic equations. Ukrainian Math. J. **42**, 1205–1210 (1990)
67. Romanov, A.V.: Sharp estimates for the dimension of inertial manifolds for nonlinear parabolic equations. Russian Acad. Sci. Izv. Math. **43**, 31–47 (1994)
68. Rouyer, J.: Generic properties of compact metric spaces. Topology Appl. **158**, 2140–2147 (2011)
69. Sakai, K.: Anosov maps on closed topological manifolds. J. Math. Soc. Japan **39**, 505–519 (1987)
70. Sell, G., You, Y.: Dynamics of Evolutionary Equations, Applied Mathematical Sciences, vol. 143. Springer (2002)
71. Sun, W.: The orbit-shift topological stability of Anosov maps. Acta Math. Appl. Sinica (English Ser.) **8**, 259–263 (1992)
72. Sun, W.: Periodic number for Anosov maps. Acta Math. Sinica, New Series **13**, 169–174 (1997)
73. Temam, R.: Infinite-Dimensional Dynamical Systems in Mechanics and Physics, 2nd edn. Applied Mathematical Sciences, vol. 68. Springer, New York (1997)
74. Tsegmid, N.: Some shadowing properties of the shifts on the inverse limit spaces. J. Chungcheong Math. Soc. **31**(4), 461–466 (2018)
75. Walters, P.: Anosov diffeomorphisms are topologically stable. Topology **9**, 71–78 (1970)
76. Walters, P.: On the pseudo-orbit tracing property and its relationship to stability. In: The Structure of Attractors in Dynamical Systems (Proc. Conf., North Dakota State Univ., Fargo, N.D., 1977), Lecture Notes in Math., vol. 668, pp. 231–244. Springer, Berlin (1978)
77. Wheeler, J.A.: Superspace. Analytic methods in mathematical physics. In: Sympos., Indiana Univ., Bloomington, Ind., 1968, pp. 335–378. Gordon and Breach, New York (1970)
78. Yano, K.: Topologically stable homeomorphisms of the circle. Nagoya Math. J. **79**, 145–149 (1980)
79. Zelik, S.: Inertial manifolds and finite-dimensional reduction for dissipative PDE's. Proc. R. Soc. Edinburgh **144A**, 1245–1327 (2014)
80. Zhong, C.K.,Yang, M.H., Sun, C.Y.: The existence of global attractors for the norm-to-weak continuous semigroup and application to the nonlinear reaction-diffusion equations. J. Differential Equations **223**, 367–399 (2006)

Printed in the United States
by Baker & Taylor Publisher Services